TECHNICAL PRINCIPLES OF BUILDING FOR SAFETY

Technical Principles of Building for Safety

Andrew Coburn, Richard Hughes, Antonios Pomonis
and Robin Spence

INTERMEDIATE TECHNOLOGY PUBLICATIONS 1995

Intermediate Technology Publications Ltd,
103-105 Southampton Row, London WC1B 4HH, UK

ISBN 1 85339 182 4

Typeset by Dorwyn Limited, Rowlands Castle, Hants
Printed in Great Britain by SRP, Exeter

Contents

Acknowledgements

The contents of this book have benefited greatly from the contributions of a wide range of people.

Firstly, our fellow team members at Cambridge Architectural Research, and authors of the companion volume *Communicating Building for Safety*, Dr Eric Dudley and Ane Haaland, assisted with conceptualizing the relevant technical knowledge as a series of principles.

Our other colleagues on the Building for Safety project, Dr Ian Davis, Alistair Cory and Andrew Clayton now at the International Disaster Research Centre, and Dr Yasemin Aysan, now at the International Red Cross, helped develop the approach to community-based building improvement programmes.

Many people reviewed early drafts of the book and assisted with comments and suggestions, including:

John Norton, Development Workshop
Peter Clark, World Relief, El Salvador
Professor A.S. Arya, University of Roorkee
Andrew Maskrey, ITDG, Peru
Paul Thompson, INTERTECT
Jinx Parker, Parker & Associates
Charles Boyle, Pacific Architects Limited
Barbar Mumtaz, University College, London
Jolyon Leslie, UNCHS, Afghanistan, and
Rodolfo Almeida, Architecture for Education Unit, UNESCO

Funding for the project was provided by the Overseas Development Administration of the government of the United Kingdom, and a special thanks is due to Michael Parkes, ODA adviser.

Graphic artwork was carried out by Vladimir Ladinski.

Introduction

This book is part of the Building for Safety Initiative of the Overseas Development Adminstration of the Government of the United Kingdom, which aims to bring knowledge of how to build safely to those who need that knowledge most.

This volume is one of four in the Building for Safety series. The other volumes are *Communicating Building for Safety*, which gives guidelines for methods of communicating technical information to local builders and householders, *Developing Building for Safety Programmes*, which presents guidelines for organizing safe building programmes in disaster-prone areas, and the *Building for Safety Compendium*, an annotated bibliography and information directory for safe building.

This volume presents the technical guidelines to be understood and adopted when building, or when managing or supervising building programmes, in hazard-prone areas. Its style, language and content are intended primarily for managers and planners of building programmes or those offering technical assistance to builders.

The information in this volume is designed to be adaptable for use in training exercises or public information campaigns. It emphasizes principles, which are universally applicable, rather than specific construction techniques, which vary with local materials and conditions.

Because of the emphasis on principles this

Training materials being discussed by builders in northern Pakistan.

Construction advice being given on-site in a post-earthquake reconstruction project, Yemen.

document should not be read as a technical reference manual for building designers, engineers or architects who are designing buildings to be built in hazard-prone areas.

Following the principles proposed will not necessarily lead to levels of safety comparable with those achievable by vigorous adherence to national codes of practice or design standards applicable to engineered buildings. For standards of protection, such as those laid out in earthquake engineering codes of practice or structural engineering design standards, readers are referred to a range of literature referenced in the bibliography. But the authors recognize that such standards are generally unaffordable by the ordinary citizens, who must nonetheless build their own shelter, and the volume has tried to avoid the scientific mystification which is often a feature of such documents.

This approach may be criticized by some engineers for not offering rigorous rules for protection. But there are already many technical manuals available offering such rules, and many well-qualified professional people willing to interpret them.

This document is intended to reach those who are or wish to be involved in building improvement programmes, but are unable to afford such rules, or are in other ways unable to obtain suitable professional assistance. The aim is to help them understand the hazards they are prone to, and the principles of building to resist these hazards, and hence to build more safely than would otherwise be possible.

Where possible a range of improvements are offered, with more sophisticated and costlier techniques giving increasing levels of protection. The builder and householder should be encouraged to build to the highest level of safety they can afford.

Building improvement project in cyclone-prone areas of the Philippines; training builders as they work.

The principles and ideas behind safer building techniques are often complex and difficult to understand or justify. For a programme to be successful it is necessary that the beneficiaries – i.e. the local community, the future occupiers – as well as the executors, that is development workers and builders engaged in the various stages, need to be involved, convinced and if possible made enthusiastic about the programme. For this, communication between all parties involved in the construction, repair or strengthening of a building is very important and must be continuous throughout the duration of a programme.

Finally, it is equally important to the success of a programme that the specific requirements of the local community and even of the prospective occupiers of individual buildings are taken into consideration and satisfied as far as possible as long as safety principles are not compromised. This consultation process is best if it takes place at the initial stage of planning, rather than later.

Chapter 1 Know your hazards

Hazard awareness

Know the history of hazards in your area.

The key to protecting against hazards is to know what to expect. Hazards are extreme occurrences of natural elements: winds, rain, tides, earth movements and weathering effects that are happening all around us every day. But occasionally they occur with such ferocity and on such a large scale that they can cause disastrous levels of damage if they are unexpected.

History lessons

Some parts of the world, some areas of countries, and individual locations, are known to be more prone to various hazards than others. In most cases if your area is likely to suffer a damaging natural hazard, a similar event will have occurred before—perhaps not in recent memory, perhaps only in the distant memory of the oldest people or perhaps even generations ago. But historical evidence and other analysis can show whether an area is at risk from hazards. Information on hazard potential is essential in building for safety. Knowing what hazards could occur in the lifetime of your building will help protect it.

Worldwide hazards

Some hazards are more likely to occur if you live in some parts of the world than others. The principal hazard areas for cyclones, earthquakes and tidal waves are identifiable from world maps, shown on the following pages. However, the fact that your area may not be indicated on the map does not mean it is necessarily safe and that the area is less prone to the hazard.

Regional hazards

Many other hazards depend on the region of the country — coastal areas and river valleys that might suffer flooding, mountainous areas that might suffer landslides. Many countries have maps zoning the country for earthquake hazards, wind storm hazards and perhaps locating important flood plain areas or coastal flood limits. Government meteorological and geological agencies are likely to have such maps. National building codes or professional engineering organizations may also have information on earthquake or wind hazards at particular locations.

Local hazards

Some hazards are more local; the threat of rockfalls, landslides, flash floods and mudflows are smaller in scale. The hazard potential can in many cases be ascertained from the layout of the land, the slopes, materials and drainage of the site. Assessing the local hazard risk from ground instabilities is a complex and skilled task, but some general indications for choosing a safe site locally are included in Chapter 2.

In the following pages, general information on hazards is presented to forewarn and forearm building improvement programme workers.

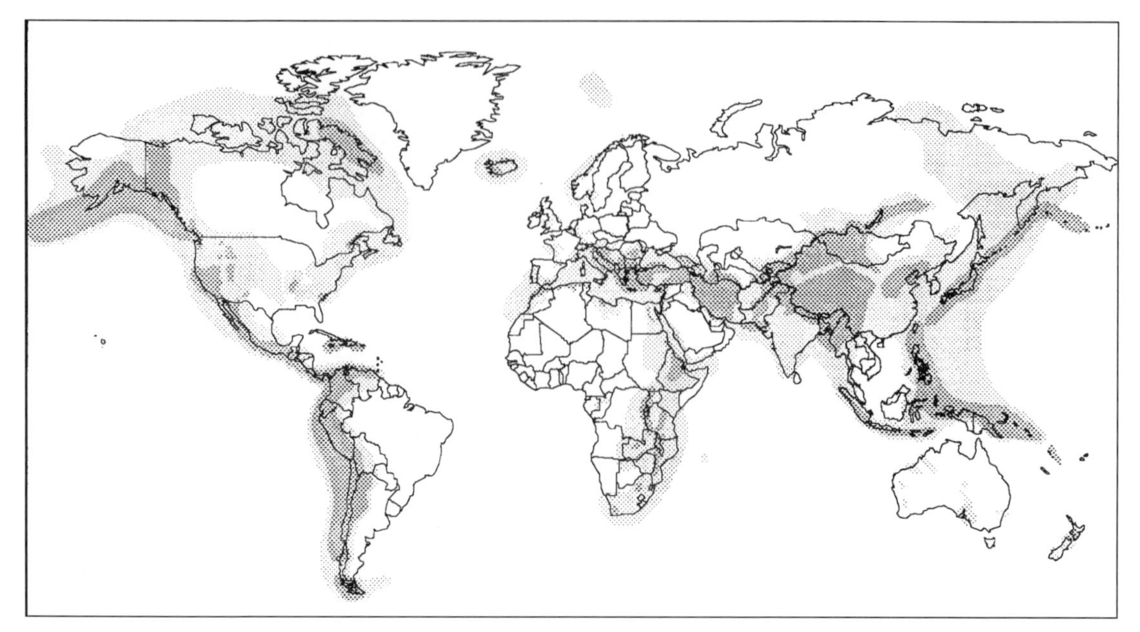

Figure 1.1 Earthquake-prone areas of the world.

Earthquakes

What causes earthquakes?

Geological adjustments deep in the earth release energy that makes the ground shake for miles around. This ground shaking is an earthquake. Large earthquakes can last for a minute or more, but most earthquakes only last a few seconds. Large earthquakes are caused by the gradual drift of continents relative to one another. Smaller earthquakes can also be caused by local geological and volcanic activity.

How do earthquakes cause damage?

Vibrational energy is transmitted through the earth causing strong shaking of the ground. Shaking causes damage and collapse of structures, which in turn may kill or injure occupants. Ground vibrations may also cause landslides, rockfalls and other ground failures, damaging settlements in the vicinity. Vibration may also overturn equipment and furniture in the home and cause damage and fires.

Where do earthquakes happen?

Maps of past occurrence of earthquakes show that they mainly occur along zones or fault systems that run through regions of countries. Earthquakes tend to recur in these areas over periods of tens or hundreds of years. Moderate earthquakes can occur almost anywhere in the world. Buildings even hundreds of kilometres away from the occurrence of a large earthquake can be damaged by the vibrations. Maps of earthquake zones in your country are likely to be available from geological agencies or included in national building codes.

What warning do you have?

Earthquakes happen suddenly and without warning. Seismologists are continuously working towards improving our understanding of the earthquake occurrence phenomenon and its observation throughout the globe. As a result we now know with reasonable confidence the areas where earthquakes are more likely to occur and their expected magnitude. However, knowledge on the time of their occurrence and the exact location of their epicentre is extremely limited. Therefore, it can be said that it is not currently possible to predict an earthquake with any accuracy.

What is most at risk from earthquakes?

Low-strength masonry buildings suffer worst in earthquakes and so do structures with heavy roofs and little lateral strength. Buildings with heavy roofs also tend to be more badly damaged. Heavier buildings also cause more injury to their occupants if they collapse. Reinforced concrete structures that are wrongly designed or poorly constructed are also quite vulnerable, and in case of collapse very lethal because of their heavy weight. Timber-framed buildings and lightweight buildings survive better if they are well-built, the elements are well-connected and the building is kept in good condition. Buildings built on loose soils, or sited on weak slopes are also at risk in an earthquake if the ground gives way. Buildings on the coast can also be subject to large sea-waves caused by undersea earthquakes.

What is the best protection?

Good strong construction is the best protection against earthquakes. Where you site your building or settlement is also important, avoiding unstable sites is the key issue (see Chapter 2).

In this book, Chapters 3 to 6 explain the principles of strong construction to resist earthquakes in buildings of different types. The earthquake-resistant construction principles detailed here are intended for small-scale, self-built houses that may fall outside the scope of seismic design codes. If you have a choice of structural materials, building in timber or using a lightweight framed system (Chapter 5) will reduce the chances of damage in an earthquake compared to building in masonry. Weak masonry (earth or rubble construction) is extremely vulnerable to earthquakes and although it can be strengthened and considerably improved (see Chapter 4), using an alternative construction material is likely to result in a safer structure.

Lighter roofs make buildings less vulnerable to earthquakes and Chapter 7 presents guidelines for safe roof construction in earthquake areas.

Care in siting a building can also reduce the risk of damage in an earthquake, particularly avoiding locations where ground failures may be triggered by the vibrations. Chapter 3 deals with siting to reduce the intensity of earthquake vibration as well as the danger of land instabilities likely to be triggered by an earthquake.

Further reading on earthquakes

Benyon, J. (1990). 'Earthquakes and Traditional Asian Buildings', *Mimar* No. 37.

Building Research Establishment (1972). *Building in Earthquake Areas,* Overseas Building Notes, 143, April 1972.

Bolt, B.A. (1988). *Earthquakes*, W.H. Freeman & Co, New York.

Coburn, A.W., and Spence, R.J.S (1992). *Earthquake Protection*, John Wiley & Sons, Chichester, New York, Brisbane, Toronto, Singapore.

Gere, J.M., and Shah, H.C. (1984). *Terra Non-firma: understanding and preparing for earthquakes*, W.H. Freeman & Co, New York.

Hodgson, J.H. (1964). *Earthquakes and Earth Structure*, Prentice-Hall, Englewood Cliffs, New Jersey.

Lagorio, H.J. (1990). *Earthquakes: an architects guide to non-structural seismic hazards*, John Wiley & Sons, Chichester, New York, Brisbane, Toronto, Singapore.

Yanev, P. (1984). *Peace of Mind in Earthquake Country: how to save your home and life*, Chronicle Books, San Francisco.

UNESCO (1987). 'Protection of Educational Buildings Against Earthquakes', by Professor A.S. Arya. *Educational Building Report* 13. Unesco Principal Regional Office for Asia and the Pacific, Bangkok, Thailand.

Wood, R.M. (1986). *Earthquakes and Volcanoes*, Mitchell Beazley, London.

Wind storms

What causes wind storms?

Winds are generated by pressure differences in weather systems.

Cyclones, alias hurricanes and typhoons

The strongest winds are generated in cyclones which occur in the tropics as severe low pressure systems several hundreds of kilometres in diameter. Cyclones are known as typhoons in the Pacific and as hurricanes in the Americas and elsewhere.

Cyclones have peak wind velocities ranging between 115 and 300km/h. The enormous pressure difference between the centre of the cyclone and its immediate surroundings is the cause of the creation of the violent winds.

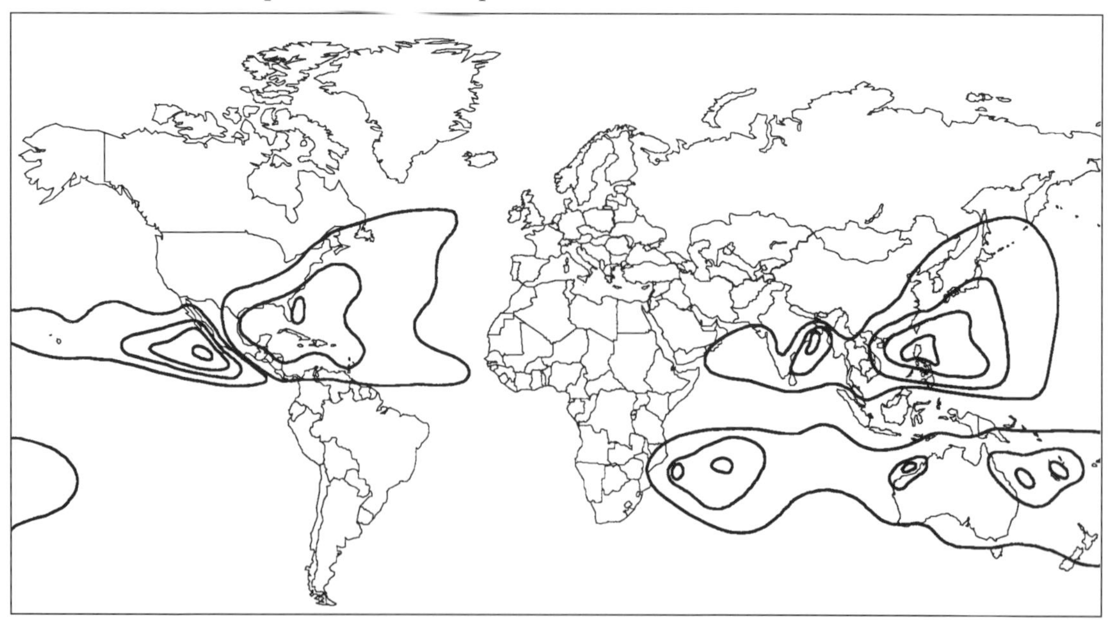

Figure 1.2 Cyclone-prone areas of the world.

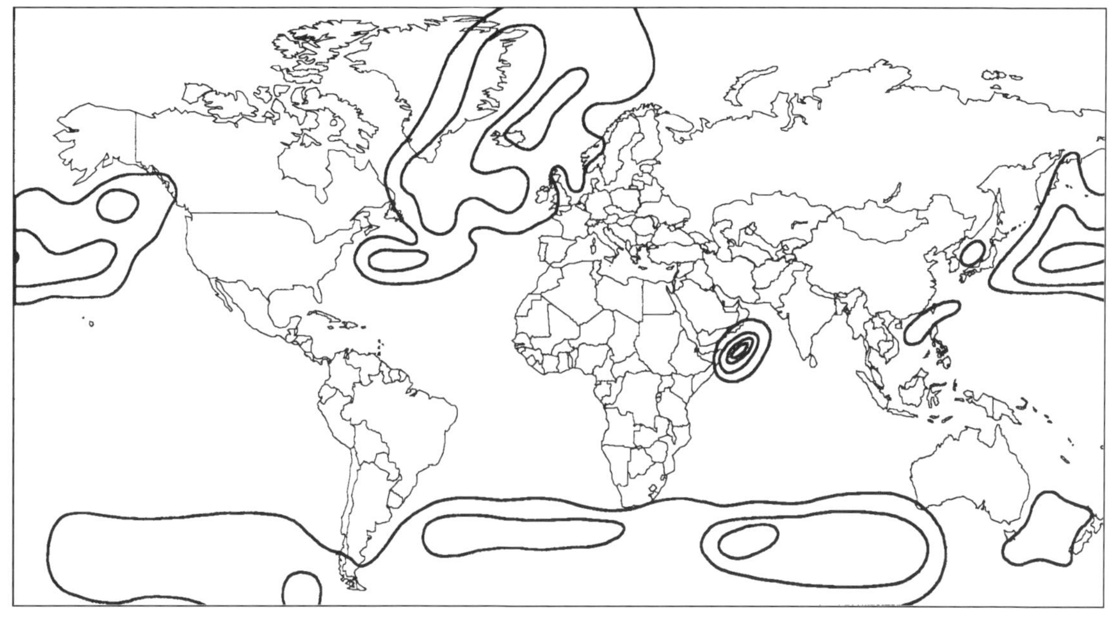

Figure 1.3 Areas of the world subject to windstorms.

Figure 1.2 shows the cyclone-prone areas of the world.

Tornadoes

Extreme low pressure areas of much narrower diameter generate rapidly twisting winds in tornadoes. This very low pressure at the centre of a tornado is extremely destructive and houses may explode on contact. The winds suck up dust and material high into the air.

Tornados are rare, occur only in certain parts of the world, and affect very localized areas.

Storms and Gales

High winds can occur almost anywhere in the world. Winds of more than 60 km/h (17 m/s) are classified as gales and can cause damage to weak buildings or poorly fastened roof materials. Winds of more than 90km/h (25m/s) are classified as storms and may cause damage to better-built structures.

Storms and gales are caused by interaction between cold and warm air-masses, powered by their thermal contrast. Large temperature differences produce turbulent disturbances with high wind speeds that can continue for several days. Figure 1.3 shows the incidence of winter wind storms across the globe.

How do wind storms cause damage?

Wind pressure

Pressure and suction from wind pressure can buffet for hours at a time. Strong wind loads imposed on a structure may cause it to collapse, particularly after hours of buffeting. More commonly, building and non-structural elements (roof sheets, cladding, chimneys) are blown loose. When the wind comes into contact with a building, in addition to the direct pressure it creates suction forces to the corners, edges, roof and around the walls, including the rear. The suction forces can be much stronger than the wind pressure.

These suction forces can rip off cladding and roof coverings, suck out infill walls, break glass windows and test structural connections to their limit.

Roofs are particularly vulnerable to high winds. (Chapter 7 deals with roof construction to resist wind damage). Damage to roofs is followed by the creation of internal forces that lead to additional damage. Damaged roofs also let in the rain and allow other damage to the building contents.

Debris impact damage

Parts of the building, particularly insufficiently tied roof sheets, and other pieces of debris carried in the air by winds become missiles that cause damage and injury on impact. Failure of non-structural elements like chimneys, towers and gable walls damage buildings.

Heavy rains

Cyclones bring with them torrential rains, and thunderstorms which can cause flooding, land slides, soil erosion and other damage to the built environment.

Sea surges

The very low pressure of cyclones causes higher sea levels. Cyclones are formed over the ocean and when they move inland, they bring with them sea surges that cause flooding across coastal areas.

Where do wind storms occur?

Meteorological records of wind speeds and direction at weather stations give probability of high winds in any region. Local factors of topography, vegetation and urbanization may affect microclimate. Past records of cyclone and tornado paths give common patterns of occurrence for damaging wind systems.

What warning do you get?

Cyclone warning systems are in operation in most affected areas of the world. Tornadoes may strike suddenly, but most strong winds build up strength over a number of hours. Low-pressure systems and cyclone development can be detected hours or days before damaging winds affect populations. Satellite tracking can help track movement of cyclones and project likely paths. Weather systems are, however, complex and still difficult to predict accurately. Today the average error in the 24-hour forecast position is around 150-200km.

What is most at risk from wind storms?

Lightweight structures and timber housing usually suffer the worst damage in wind storms. Informal makeshift squatter settlements are often badly affected. Roofs and cladding sheets are the most vulnerable parts of structures. Sheets, boards or poorly fixed building elements may be pulled loose. Large or old trees are vulnerable to strong winds. Damage is also commonly inflicted on fences, billboards and other street items. Boats and fishing or other maritime industries tend to suffer badly if caught at sea.

What is the best protection against windstorms?

The best protection against windstorms is to build strong structures capable of resisting their forces and to connect all structural and non-structural elements together. Regular maintenance is also very important.

Strong buildings

Chapters 3 to 6 describe measures for making different types of building robust and resistant to the pressures of winds and other hazards.

Wind-resistant roofs

It is particularly important to build a strong roof and to fasten it firmly to the building structure. Chapter 7 describes roof construction for wind and earthquake resistance and gives some maintenance principles.

Sheltered siting

Siting of buildings and settlements can also contribute to their safety, by finding locations less exposed to the wind or protected by windbreaks. Chapter 2 deals with siting to reduce exposure to wind and also the danger of land instabilities and flooding that might be induced by wind storms.

Cyclone shelters

In some communities, an additional protection measure employed is the construction of especially strong communal shelters. These are capable of accommodating the population of the vicinity and engineered to keep them safe from the wind damage, floods and other threats likely to affect the rest of the buildings. When cyclone warnings are broadcast, everyone is encouraged to take refuge in the cyclone shelter until the storm has passed.

In other areas, each householder has been encouraged to build a cyclone-resistant core or extension to his or her own house. This small room, built of strong masonry, reinforced concrete or firmly braced timber frame acts as a shelter and store to protect from the cyclone.

Further reading on wind storms

Asian Development Bank (1991). *Disaster Mitigation in Asia and the Pacific*, available from: Information Office, ADB, P.O. Box 789, 1099 Manila, Philippines.

Bodschalk, D., Brower, D., and Beatley T. (1989). *Catastrophic Coastal Storms: hazard mitigation and development management*. Duke University Press, Durham, North Carolina, USA.

Buller, P.S.J. (1986). *Gale Damage to Buildings in the UK – an Illustrated Review*, Building Research Establishment, Watford, UK.

Diacon, D. (1992). *Typhoon-resistant Housing in the Philippines — A Success Story.* Information from: Building and Social Housing Foundation, Memorial Square, Coalville, Leicestershire LE6 4EU, UK.

Eaton, K.E. (1980). *How to Make Your Building Withstand Strong Winds*, BRE Public.

Eaton, K.E. (1981). *Buildings and Tropical Windstorms*, BRE Overseas Building Notes, 188.

Eaton, K.E. and Reardon, G. (1985). *Cyclone Housing in Tonga*, BRE Public.

Fairbridge, R.W., (ed.) (1967). 'Tornadoes and hurricanes', in *The Encyclopedia of Atmospheric Sciences*, 1003-6. Downden, Hutchinson & Ross, Stroudsburg, Pennsylvania, USA.

Greenwood, R.F. (1992). *Hurricane-resistant Construction*, The Reporter Press, 147 West Street, Belize City, Belize, Central America.

Holthouse, H. (1986). *Cyclone: a century of cyclonic destruction*, Angus and Robertson, Sydney.

Mayo, A. (1988). *Cyclone-resistant Housing for Developing Countries*, BRE publication distributed by Intermediate Technology Publications.

Munich Re (1990). *Windstorms*, p.120, Available from: Munich Re, Koniginstrasse 107, D-8000 Munchen 40, Germany. Order No. 1672-V-e.

Nalivkin, D.V. (1983). *Hurricanes, Storms and Tornadoes*, Balkema, Rotterdam.

Norton, J. G. Chantry, and Ngyen Si Vien (1990). *Typhoon-resistant Building in Vietnam*, in *Mimar* No. 37.

Pielke, R.A. (1990). *The Hurricane*, Routledge, London.

Shri Reddy, I.A.S. (1992). *Building for Safety: a case study of cyclone-prone coastal region of the eastern state of Andhra Pradesh, India.* Information from: National Institute of Rural Development, Rajendranagar, Hyderabad-500 030, India.

Simpson, R.H., and Riehl, H. (1981). *The Hurricane and its Impact*, Blackwell, Oxford.

UNESCO (1987). *Typhoon-resistant School Buildings for Vietnam*, Bangkok, July 1987. Available from: UNESCO, Educational Architecture Unit, 7 Place de Fontenoy, 75700 Paris, France.

UNDRO, (1978). *Disaster Prevention and Mitigation*, Vol. 4 Meteorological Aspects. New York 1978. Available from: Office of the United Nations Disaster Relief Coordinator, Palais de Nations, CH-1211, Geneva-10, Switzerland.

Whipple, A. (1982). *Storm*, Time-Life Books, Amsterdam.

Floods

What causes floods?

There are four main types of flooding:

1. ***Coastal flooding,*** where high tides, large waves and wind storm surges bring sea water further inland than usual. Cyclones (hurricanes and typhoons) which are very low pressure weather systems, cause higher sea levels and are often causes of coastal flooding.

Another special type of coastal flooding is a *tsunami*—a freak height tidal wave caused by earthquakes out to sea.

2. ***River flooding,*** where rivers carrying more water than usual burst their banks—this is more likely with large rivers on flat lands. Rivers draining a large geographical catchment area can become overloaded when

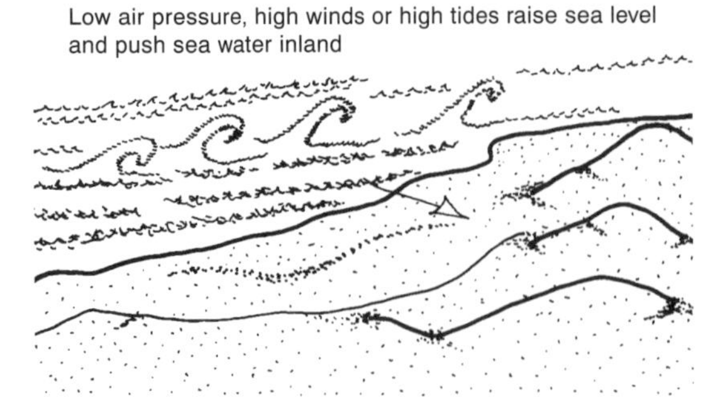

Figure 1.4 Coastal flooding

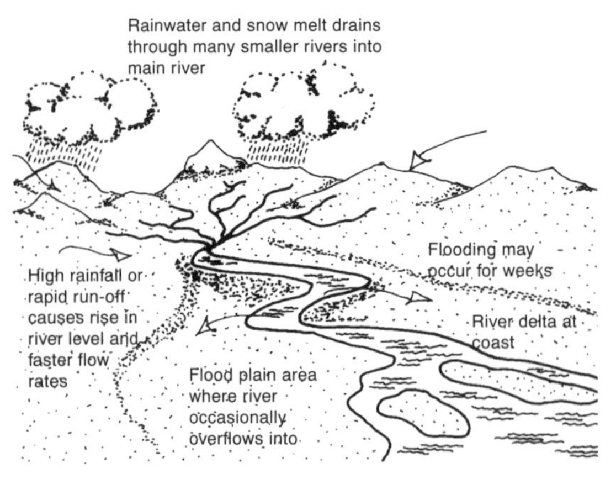

Figure 1.5 River flooding

exceptional rainfall or rapid snow-melt occurs, perhaps many kilometres upstream. The loss of forests and soil cover in the rain catchment area may also increase the speed of rainwater runoff into the river system and make river flooding more likely downstream. Upstream dams if allowed to silt up lose a large part of their capacity and water might have to be released periodically. This is increasingly an additional cause of flooding.

3. ***Flash flooding***, where sudden releases of water, such as after exceptionally heavy rainfall or a lake overflow, are channelled downstream. This is a threat in mountainous areas and where rainwater runoff comes quickly or is constrained by steep valley-sides, deep gulleys or through ravines.

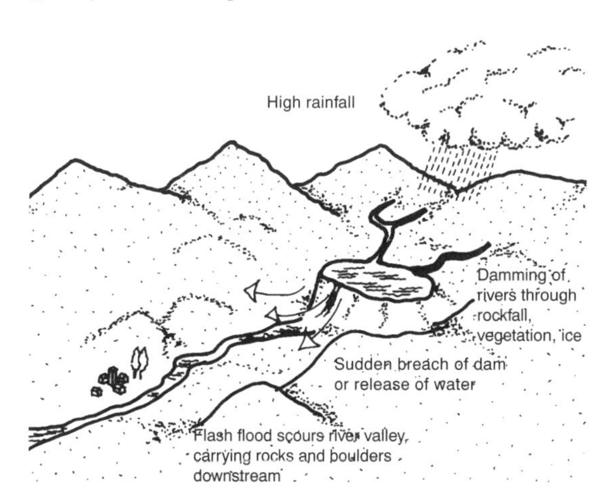

Figure 1.6 Flash Flooding

4. ***Urban flooding***, where heavy rainfall does not drain away quickly enough through the hard landscape of towns. Undrained rainfall can quickly form artificial rivers in the streets, flooding into buildings and causing damage and disruption. Densely urbanized areas with a high proportion of the ground built on or paved over and with insufficient drainage systems are particularly prone. Urban flooding can be made worse by debris clogging up drainage channels. In many cases river channels have become dumping sites for large unwanted items or even refuse and scrap leftovers from neighbouring building sites.

How do floods cause damage?

Currents of moving or turbulent water can knock down and drown people and animals in relatively shallow depths. Moving water is powerful and destructive; the faster it flows, the more destructive it is. The depth and velocity of the water determine how much damage a flood causes. Debris carried by the water is also destructive to people, property and buildings. Mud, oil and other pollutants carried by the water are deposited and will destroy crops and building contents. Flooding destroys sewerage systems, pollutes water supply and may spread disease. Saturation of soils may cause landslides or ground failure.

Where do floods happen?

Floods most often happen where they have happened before. Historical records give a first indication of how often floods occur (their *return period*), how far they extend and how high or severe they are likely to be. Many official meteorological agencies or river authorities keep maps of past floods or areas most at risk from flooding. Topographic mapping and height contouring around river systems and low-lying coastal areas can also give indications of where river and coastal flooding might occur if historical records are not available. Many other factors affect the likelihood of flooding, including the capacity of the hydrological system and rainfall catchment areas, precipitation and snow-melt records, tidal records, storm frequency, coastal geography and other features.

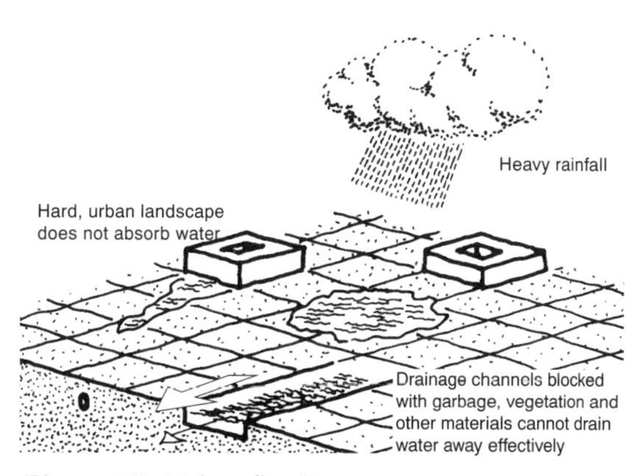

Figure 1.7 Urban flooding.

How can floods be prevented?

Most flood prevention measures require large civil engineering works to regulate and contain the powerful flood waters. These measures may include retaining walls and levees along rivers, sea walls along coasts, storm drainage, river dredging, polder systems and building other flood control devices. Water regulation (slowing up the rate at which water is discharged from catchment areas) can be achieved through construction of reservoirs, increasing vegetation cover to slow down runoff, and building sluice systems. Alternative drainage routes (new river channels, pipe systems) may prevent river overload. Storm drains in towns assist drainage. Beaches, dune belts and breakwaters also reduce the power of tidal surges.

What warning do you get?

Flooding may happen gradually, building up depth over several hours, or suddenly with flash flooding. Heavy prolonged rainfall may warn of coming river flood or urban drainage overload. High tides with high winds may indicate the chance of coastal flooding some hours before it occurs. Cyclones likely to cause flooding are usually monitored and official agencies can often predict their arrival a day or two in advance. Where sufficient warning of flooding is given, evacuation may be possible if a proper preparedness plan has been organized beforehand. It is usually difficult to improvize an effective evacuation programme without preparing beforehand.

What is most at risk from floods?

Anything sited in flood plain areas is at risk from floods. Buildings most likely to suffer structural damage in contact with flood waters include earth buildings or masonry with water-soluble mortar. Buildings with shallow foundations, older or poorly-built structures with weak resistance to lateral loads or impact are also likely to be damaged. Basements or underground buildings will be completely flooded. Utilities are easily disrupted by floods, particularly sewerage, power and water supply. Building contents likely to be spoiled by contact with flood waters include all soft furnishings, food stocks, machines and electronics. Livestock, particularly confined or penned, tend to suffer high levels of loss in floods.

Tsunamis

A special type of coastal flooding is the tidal

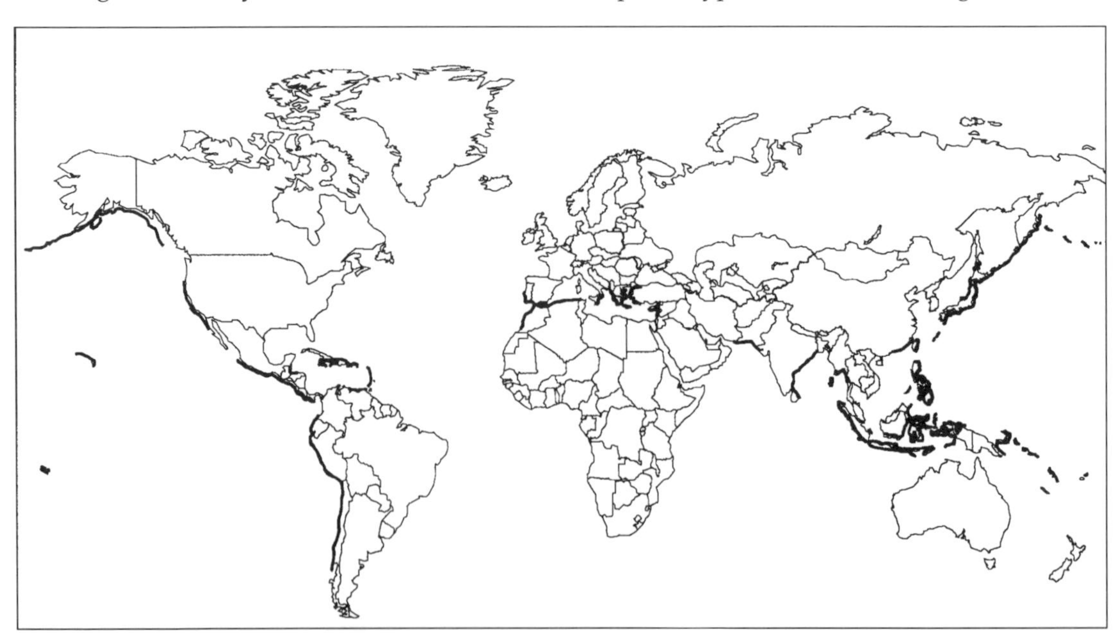

Figure 1.8 Coastlines that may experience tsunamis

wave caused by large earthquakes or volcanoes out to sea. These can cause sea waves that strike coastal areas with extreme force. The speed with which they approach the coast can exceed 300 kilometres an hour, and as they strike the coastal areas they can reach heights of tens of metres. Such waves are rare but destructive when they occur. *Tsunamis* have affected coasts in the Pacific, the Caribbean and Mediterranean sea, as indicated in Figure 1.8.

What is the best protection?

The best protection against floods is careful siting to avoid building in the worst flood-prone areas.

Siting for flood protection

Chapter 2 presents siting principles to help discover the likelihood of flooding in a particular site and selection of sites to minimize risk, both from floodwater and land instabilities likely to be triggered by it.

Construction in flood-prone areas

In areas where flooding is a possibility, the construction of earthen or rubble building types is inadvisable, as the structural materials themselves are vulnerable to water and almost impossible to protect. Other building types can be constructed strongly enough to withstand moderate pressures from floodwater without suffering serious structural damage if the principles of robust design, good foundations and strong construction are followed, as presented in Chapters 3, 5 and 6.

Wall base protection

Additional protection against damage from floodwater include using concrete, rendering or buttressing at the base of walls to protect them from floodwater erosion or impact of water-borne debris. Raising the ground floor level of buildings makes it less likely that floodwater will enter and damage contents. Any solid plinth or columns used to raise the floor level need to have firm foundations and strong horizontal bracing to prevent damage from lateral forces.

Further reading on floods

Cuny, F.C. (1991). 'Living with floods: alternatives for riverine flood mitigation'. In *Managing Natural Disasters and the Environment*, Kreimer, A., and M. Munasinghe (ed.), Environment Department, The World Bank.

FEMA (1981). *Design Guidelines for Flood Damage Reduction*, United States Federal Emergency Management Agency, Washington DC.

FEMA (1984). *Design and Construction Manual for Residential Buildings in Coastal High Hazard Areas*, United States Federal Emergency Management Agency, Washington DC.

Handmer, J. (ed.) (1987). *Flood Hazard Management: British and international perspectives*, Norwich, England: Geobooks.

Irrigation Support Project for Asia and the Near East, ISPAN (1992). *Flood Proofing Study*. Bangladesh Ministry of Irrigation, Water Development and Flood Control.

Parker, D.J., Green, C.H., Thompson, P.M. (1987). *Urban Flood Protection Benefits: a project appraisal guide*, Technical Press, Franborough, UK.

United Nations (1976). *Guidelines for Flood Loss Prevention and Management in Developing Countries*, United Nations Department of Economics and Social Affairs, Natural Resource Water Series No. 5.

Ward, R.C. (1978). *Floods: a geographical perspective*, Macmillan, New York.

Land instabilities

What causes landslides?

Landslides are the collapse of sloping soils or rocks. Rain, wind or other erosion forces gradually break down the steep slopes of mountain sides. Earthquake shaking is another cause of landslides in areas of unstable slopes. The steeper the slope, the less stable it is likely to be.

There are several types of land instabilities:

Landslides

When gravitational forces imposed on sloping soils exceed the shear strength or sliding resistance of soils that hold them in position, soils slide downhill. High water content makes soils heavier, due to the increase in the loads. It also results in a decrease in the shear strength of the soil, that becomes weaker. So heavy rainfalls or flooding make landslides more likely to happen. The angle of slope at which soils are stable is a material property of the soil. Steep cuttings through some types of soil makes them unstable. Triggering of the collapse of unstable soils can be caused by almost any minor event: storms, minor ground tremors or man-made actions. Building activity changes the drainage and profile of the site and may result in an increased risk.

Rockfalls

Rockfalls are common in unvegetated, very steep rock cliff-faces. The majority of rockfalls occur during earthquakes or heavy rainfalls. Rock cliff-faces usually develop in resistant bedrocks that do not weather or slip easily (eg sandstone, limestone and so on). These cliff-faces are usually supported by debris and erosion material at their base.

Mud and debris flows

A flow is any slope failure where water is a big part of the sliding material and the major factor in its behaviour. When the flow is dominated by clayish materials then it is called a mud flow or earth flow or mudslide. If the range in the particle size of the materials of the flow is highly varied with smaller and larger particles then the flow is called a debris flow.

How do landslides cause damage?

Landslides destroy structures, roads, pipes and cables either by the ground moving out from beneath them or by burying them. Gradual ground movement causes tilted, unusable buildings, cracks in the ground split foundations and rupture buried services. Sudden slope failures can take the ground out from under settlements and throw them down hillsides. Rockfalls are extremely dangerous to buildings because boulders that roll down and collide with structures and settlements are very destructive. Debris flows cause destruction when slurry material, man-made spoil heaps and soils with high water content flow like a liquid, filling valleys, burying settlements, blocking rivers and roads (possibly causing floods).

How can land instabilities be identified?

Previous landslides or ground failures can be identified by geotechnical surveys. Fissures in the ground, certain characteristics of the shape of the slope, tilting trees and distorted surface features may all indicate that ground movement has occurred or is still happening slowly. Steep slopes and soft soils may also indicate risk of a landslide in the future.

Mapping of soil types (surface geology) and slope angles (topographic contouring) can identify where slope failures are most likely to occur across a region. In addition, mapping of water tables, hydrology and drainage also help to identify site risks. Other likely candidates for unstable slopes include artificial landfill, man-made mounds, garbage tips and slag heaps.

Can landslides be prevented?

Major landslides are almost impossible to prevent or contain. Where a serious landslide has occurred, or is possible, the area should not be built on. For less severe cases, the landslide risk of a slope can be decreased by reducing the slope angles, either by terracing the slopes or excavating the slope back. Other measures to reduce the chances of slopes failing include increasing drainage of the slope (both deep drainage and surface runoff) and engineering works (eg piling, ground anchors, reinforced earth and retaining walls). When building roads through hilly terrain, the angle of cuttings and embankments should be as shallow as possible to reduce risk of slope failures.

Some stabilization of slopes and erosion limitation is possible by planting trees. Forestation can prevent the loss of surface material down to the depth of root penetration. Roots help to bind surface planting and anchor the soil. Indigenous plants with strong roots and rapid growth are often a good remedy.

Where the primary hazard is material falling down on to settlements, protection barriers can be built. These include trenches, slit dams and vegetation barriers of specially planted trees and bushes.

Where predictable down-washes of debris occur, debris flows can be directed into specially constructed channels.

What is most at risk from landslides?

Settlements built on steep slopes and softer soils are most prone to ground failure. Sites along cliff tops may be at risk from cliff erosion. Settlements built at the base of steep slopes, on alluvial outwash fans or at the mouth of streams emerging from mountain valleys may be at risk from rockfalls or debris flows. Roads and other communication lines through mountain areas, if not properly drained and designed, may be interrupted by slope failures. Buildings of low-strength or brittle masonry, and those with weak foundations can suffer subsidence damage which weakens their structure.

What is the best protection against landslides?

The best protection is to avoid using areas potentially likely to be affected by slope failures for settlements or as sites for important structures. Chapter 2 deals with site appraisal for landslide risk. Structures built on slopes should be designed to withstand or accommodate potential ground movement. Well-connected foundations on larger structures will protect them from being affected by differential settlements. Where buried services cross slopes, they should be designed to be flexible.

Further reading on land instabilities

Anderson, M.G., Richards, K.S., (ed.) (1987). *Slope Stability: geotechnical engineering and geomorphology*, John Wiley, New York.

Blair, M., *et al.* (1984). *When the Ground Fails: planning and engineering response to debris flows*, Natural Hazards Research and Applications Information Center, Boulder, Colorado, USA.

Bromhead, E.N., (ed.) (1986). *The Stability of Slopes*, Chapman & Hall, London.

Brunsden, D., Prior, D.B. (1984). *Slope Instability*, John Wiley, Chichester, England.

Crozier, M.J., (1986). *Landslides: causes, consequences and environment*, Croom Helm, London.

Gray, D.H. and Leiser, AT (1982) *Biotechnical Slope Protection and Erosion Control,* Van Nostrand Reinhold Co, Holland.

Veder, C. (1981). *Landslides and their Stabilization*, Springer, New York.

Záruba, Q., Menci, V. (1982). *Landslides and their Control*, Elsevier, Amsterdam.

Chapter 2 Choosing a safe site

Siting and safety

Building safely begins by choosing a safe site. A lot of protection against hazards can be gained by careful location. For some hazards, like flooding or land instability, siting is the most important protection measure. For other hazards, like earthquakes or wind storms, where a building is sited is less critical than how well it is built, but location can still be important in determining how strongly the hazard is experienced and the potential for follow-on hazards causing damage.

The degree of choice that people have in siting their structures is highly variable: some may have little or no choice, others may have a large area within which to choose a site. In this chapter, the hazard considerations of a site are outlined so that those with some freedom of choice can exercise it to protect themselves to the maximum extent possible. Possible modifications to existing sites to reduce their hazard potential are presented.

Siting can be considered on a number of scales: at a community scale where perhaps there is a building improvement programme to locate or relocate a community within a geographical area, and at an individual scale when choosing a site for a building within a locality.

Siting a settlement

Safe siting for a new settlement

In locating a new town, village or housing estate, the hazard context is only one of the many factors that should be taken into account to make a successful settlement. It should therefore be considered at an early stage of site selection. The penalties of siting on ground with possible hazard problems should be balanced against the advantages of locating there for other reasons.

Flat land that makes larger settlements more economical is likely to be found in valley floors, river plains or coastal areas that could conceivably be flood-prone.

Steeper sites may incur other problems, such as potential landslides, rockfalls or, because foundation construction is more difficult, may cause building settlement and other difficulties.

When a large site is needed, it is possible that some areas within the land chosen will be safer than others. Therefore, sensitivity to the use of the land within the settlement will be important.

In choosing a site for a new settlement, consider the topography, geology and soils, the climate, vegetation, and current land use. Each of these will give clues to the hazards prevailing on the site and may suggest ways of siting with maximum regard for safety. Watershed areas should be kept clear of settlements for environmental reasons.

Relocation after disaster

When a community has suffered heavy damage from a natural hazard, it is not uncommon for serious consideration to be given to relocating the community as a way of reducing future hazards. Where a community suffers frequent damage, or the site itself is fundamentally unsafe, this may be the only practical option.

However, where damage has been inflicted by a rare event, such as an unusual flood or an earthquake, then resettlement to a new site should be resisted. Relocating a settlement to reduce future hazards is rarely popular with the occupants and, as a lot of experience has shown, usually results in a poorer community in the longer term. Resiting can have a severe effect on the local economy by changing agricultural conditions, microclimate and water supply. The loss of infrastructure, land ownership and building stock, even if badly damaged, is rarely recouped by starting again on a fresh site.

The decision on whether to relocate a damaged settlement is a difficult one and should be very carefully considered. The capital investment needed to turn any relocation into a success is considerable.

Relocation of a settlement should not be carried out without the full support of the community and without understanding the reasons why the new settlement was originally sited where it was, and without convincing reasons that the new site is significantly better.

Improving the safety of an existing site

Most settlements are already well-established. A community concerned about its hazard potential may well consider reducing the risk by taking steps to improve the safety of the site as a whole. Some safety measures can only be taken by the entire community at risk as they are beyond the capability of each individual householder to carry out independently.

The community as a group may be able, for example to terrace slopes for safer building, or improve drainage to reduce the risk of saturated soils causing landslips.

Community efforts can build levees around their villages and towns to protect against potential floodwaters.

Planting trees as windbreaks can reduce the community's exposure to wind damage and communal tree planting can also be used to stabilize the topsoils of slopes where they are in danger of being washed away. Trees and shrubbery can be planted as barriers against potential rockfall routes.

The community as a group and through its representatives can decide the priorities of siting an individual building—perhaps agreeing to site its school or communal buildings in the safest locations.

Perhaps one of the most important roles for the community as a whole is to be aware of the importance of siting and building safely; to discourage members of their community from siting on the most hazardous areas; to encourage their neighbours to protect themselves; making sure that younger generations of the community grow up understanding the hazards around them and building safely for the future.

Siting a building

Finding a safe site for a building is very important. The choice of site available may be broad or it may be very limited. Either way it is important to understand the hazard context of the available sites and to choose the best. It is important to look for evidence of hazards in a broad area around the candidate sites. When a site has been chosen its immediate surroundings and the building site itself must be examined further and more carefully. Often such evidence is difficult to prove without the help and involvement of a specialist. In such a case the advice of a specialist might be necessary. Local knowledge about the sites must always be taken into consideration: beware of land that has traditionally been avoided for building on.

Understanding the siting considerations for various hazards is discussed in later sections of this chapter.

In addition to avoiding specifically hazardous features, a number of siting considerations are important.

Dig into the site

Carry out a site investigation: dig down into the soils to some depth to understand what lies below the immediate topsoil. Foundations should be founded at a depth in the ground that it is hard to dig further with pick axe and shovel. The water table is important—it should be normally well below foundation levels.

Firm ground

Good firm soils—rock where it is available—are the best sites for strong foundations. Building a structure on strong foundations is one of the most important principles of building for safety, as recommended in Chapters 3 to 6. Softer soils make strong foundations more difficult to construct: they have to be dug deeper, made more massive and tied together horizontally to prevent the structure settling and becoming weakened. If possible, choose a site where the ground conditions are constant across the area of the site: a site that changes from firm soil in one area to soft soil in another is very difficult to build good foundations on.

Flatter sites

Steep slopes are generally hazardous. They have a higher risk of the ground slipping or of material coming down the slope. A steeply sloping site creates stability problems. Foundations should be level—stepped foundations weaken a structure. Terracing to form a flat level site on a slope or digging down into the slope to create foundations as level as possible will improve safety conditions. However, in sites that might become unstable due to their steepness, too much terracing and cutting-in due to high-density housing might destabilize the area further.

Steep sites are hazardous and create vulnerable buildings—avoid if possible. Hillside collapse, Philippines.

Drainage paths

Avoid siting the building across natural drainage paths. As well as noting obvious stream channels and drainage ditches, check the site when it is raining to identify the runoff routes. Sites which may be flooded and waterlogged should be avoided.

Improving safety of an existing building

Where a building already exists, there may be ways of improving its general safety by improving the site and surrounding areas:

- Improve drainage, particularly to prevent water seeping into foundations, or pooling at the base of walls which will weaken the structure

- Protect the building structure from being damaged by animals, vehicles or other agents. Use fences and raised ground level to prevent the area close to the walls from potentially damaging activity
- Windbreaks can be erected or planted in areas of high winds. Windbreaks also protect buildings from harsh exposure to weathering and driving rain which will make a structure more vulnerable to future hazards
- Some protection against flooding may be possible by raising the ground level, building dykes around the site or raising the door threshold height
- Keep trees and shrubs from growing too close to the building as their roots may damage the foundations or in clay soils, may cause shrinkage cracking and subsidence. Furthermore avoid building near to big trees that might fall during a storm
- Check waste and water connections to eliminate leakages, and
- Keep ground level well-below a damp proof course.

Safe siting in flood hazard areas

Higher sites are less flood-prone

Checking flood hazard

In low-lying areas close to the coast and on flat land in river valleys there may be potential for coastal or river flooding. In geologically younger river valleys in mountains and foothills there may be potential for flash-flooding.

It is important to check the history of flooding in the area. Where possible:

- Map the extent of land covered by past floodwaters
- Get an indication of the depth of past floodwaters
- Find out about the severity of past floods; how much damage they caused, how fast they flowed and how much debris they left behind, and
- Find out how often flooding has happened over at least the past 20 years.

Local knowledge

The local community will know of any flooding in the recent past. Earlier floods may be part of traditional folk tales and stories. Other local knowledge may help in establishing where flooding is likely.

Official records

There are a number of local or national authorities that may be able to help with information on past flooding occurrences. Some may map or measure flood occurrences.

Local or national authorities that may be able to help with information on past flooding occurrences in the area include:

- River and coastal authorities
- Irrigation authorities
- Maritime, oceanographic or tidal monitoring stations
- Local or regional land-use regulatory departments, and
- Weather stations.

Physical inspection of the area

Land morphology is the main factor in determining how safe a site is against floodwaters. Flood plains of rivers are typically broad, flat valleys containing a large or meandering river. Coastal flood plains can extend a long way inland from the sea itself, if the land is flat, without hills or ridges between it and the sea.

The higher the elevation above average river or sea levels, the less likely it is to be covered by floodwaters.

Sites below the level of nearby rivers or seas, for example below river banks, or in enclosed depressions are highly likely to suffer flooding at some time or another.

Other possible indicators that a site is very regularly flooded include:

- Water-logged soils
- Marshy lands, and
- water-loving vegetation (eg reeds, marsh-grass).

Signs that a site has suffered flooding at some time in the past include:

- Silty (fine-grained) soils
- River high-water marks (eg water stains, erosion notches)
- Deposits of rocks, trees or debris from elsewhere, and
- Old river courses.

The most dangerous locations in flood-prone areas include steep earth banks, gorge sides or slopes on the sides of rivers or river valleys where high floodwaters are likely to cut into and undermine the slopes.

The most dangerous locations in flood-prone areas are river banks and the sides of gorges. River edge development, Philippines.

Consider the possible implications of floods on access routes and services to your site.

Sea wave protection

In open coastal areas where storm surge, tidal waves or tsunami are a possibility, it is best if high-density housing, capital-intensive infrastructure, hazardous facilities and other important buildings, are built on higher ground or at least 250m from the coast.

Safe siting in areas with land instabilities

Building on a site safe from landslides, rockfalls mud and debris flows.

An important aspect of siting is the stability of the ground itself. Identifying which areas of ground are more safe to build on may prevent damage occurring from landslides, rockfalls or mud and debris flows.

The selection of a stable site is even more important in areas of earthquake hazard, or regions likely to suffer monsoons or heavy rainfalls because both ground vibration and heavy rainfall are likely to increase the chances of ground failures.

Recognizing the topography

Certain topographic features, like cliff edges and ravine sides, are in themselves recognizable as hazardous locations on which to build.

Slope stability

Other slopes can be assessed for their failure potential before building on them.

High failure potential

Slopes with a high failure potential include:

- Gradient more than 30°
- Slopes with all the following characteristics:
  a. Gradient more than 15°
  b. Underlying geological formation of poor or loosely consolidated material
  c. Soft, thin soil cover (indicated in the first instance by thin vegetation coverage, lack of trees, etc), and
- Slopes with a previous history of failure, or evidence of landslide, including fissured ground, significant soil downwash or visible signs of rockfalls (eg boulder-strewn lower slopes).

Moderate failure potential

Slopes with a moderate failure potential include:

- Gradient above 15°
- Slopes with all the following characteristics:
  a. Gradient above 8°
  b. Underlying geological formation of poor or loosely consolidated material
  c. Soft, thin soil cover (indicated in the first instance by thin vegetation coverage, lack of trees, etc)
- Slopes with other factors that increase the probability of failure, including:

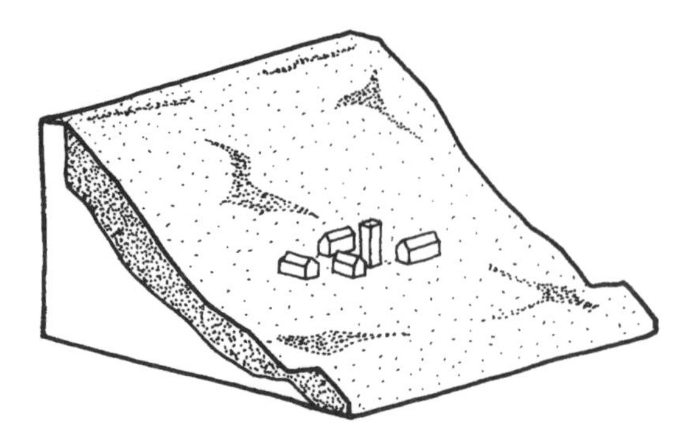

*Figure 2.1 Hazardous sites include...
Recent landslides and unstable slopes.*

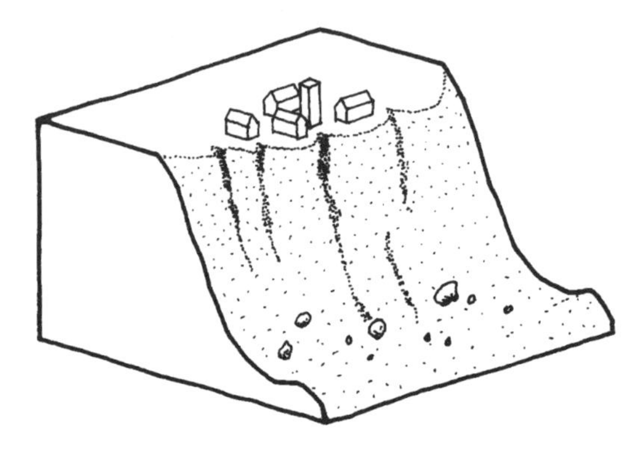

*Figure 2. 2 Hazardous sites include...
Edges of fragmentatious cliffs, scarp or terrace.*

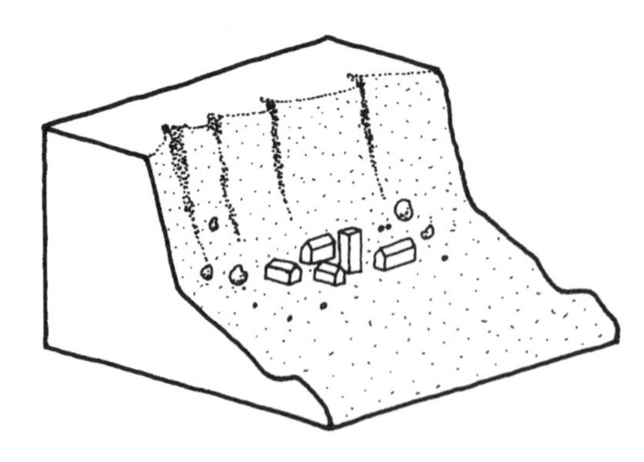

*Figure 2.3 Hazardous sites include...
Areas exposed to rockfall and debris landslides.*

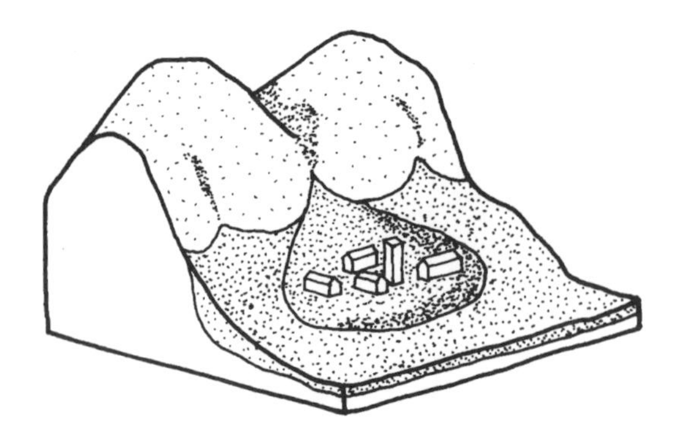

*Figure 2.4 Hazardous sites include...
Soft slope deposits or valley outwash fans.*

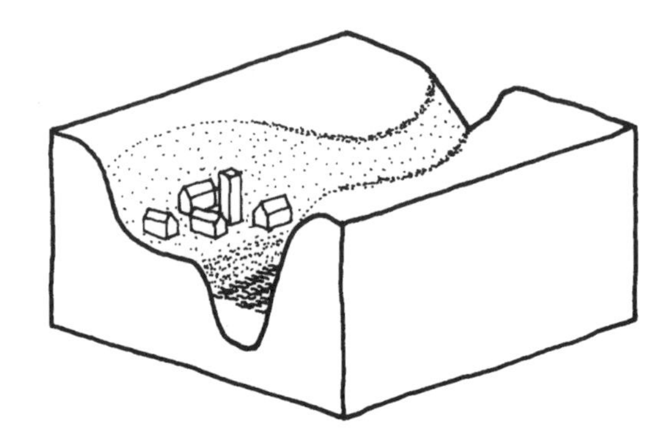

*Figure 2.5 Hazardous sites include . . .
Within a river gorge or on the edge of a water channel.*

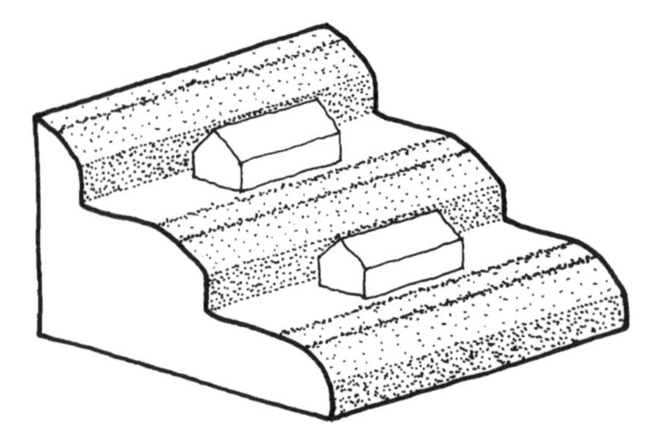

Figure 2.6 Landslide risk can be reduced by terracing slopes.

a. Outwash fans
b. Slopes down to river courses
c. Slopes with cuttings through them, and

- Steeper slopes with mitigating factors such as:
  a. Retaining walls or slope engineering
  b. Special drainage
  c. Thick tree cover.

Steep slopes, like the edges of gulleys or drainage channels can suffer from severe erosion and undermine buildings built nearby.

Stabilizing slopes

In cases where the risk of failure is not severe, the landslide risk of a slope can sometimes be decreased by reducing the slope angles, either by terracing the slopes or excavating the slope back.

Other measures to reduce the chances of slopes failing include:

- Increasing drainage of the slope (both deep drainage and surface runoff), and
- Engineering works (eg piling, ground anchors and retaining walls).

Some stabilization of slopes is possible by planting trees. Forestation can prevent loss of surface material down to the depth of root penetration. In most regions there are indigenous plants that are good in slope stabilization.

The landslide risk of a steep slope can be reduced by terracing and the construction of retaining walls.

Rockfall barriers

Where the primary hazard is material falling down on to settlements below, protection barriers can be sometimes built, including trenches, slit dams and vegetation barriers of specially planted trees and bushes.

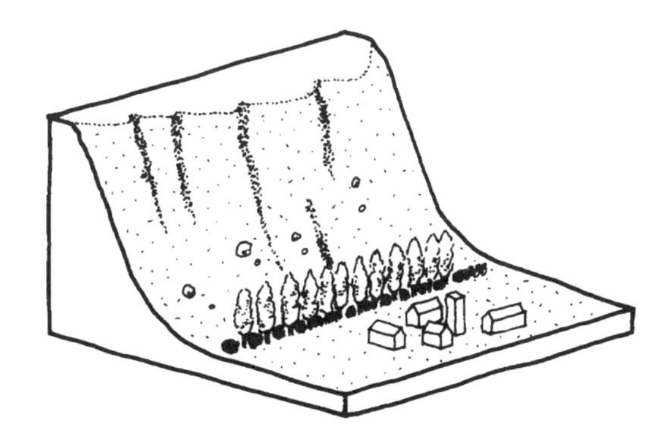

Figure 2.7 Vegetation screens can provide a protective barrier against rockfalls.

Safe siting in wind-prone areas

Find a site sheltered from the prevailing wind.

The regions of the world that are most prone to cyclones and other high winds are mapped in Figure 1.2, but there is much local variation. It is essential to know the severity of your wind hazard locally. Find out how severe past storms have been, how common they are and the predominant direction that most storm winds blow.

Cyclones have relatively short return periods — they tend to occur every few years at least, so local knowledge about them is probably reliable, although the next may be more severe. Other sources of information include local meteorological institutes, and weather stations. Some cyclone-prone countries are zoned for wind speed hazard.

Siting to reduce exposure to the wind

Certain factors make some sites more exposed than others:

- Coastal areas are particularly prone: cyclones originate out at sea and become hazardous when they come ashore. They also drive the sea level up to cause coastal flooding
- Estuaries and river deltas, that will flood during the heavy rainfall associated with the cyclone
- Exposed sites on the tops of hills, or cliff tops. Winds can be up to fifteen per cent stronger on elevated sites, and
- Valley necks or open-ended valleys, through which winds may be channelled.

When siting in areas that suffer from high winds:

- Select a sheltered site. Use any topographical effects or natural defences that may protect your building or settlement from the prevailing wind
- Consider the orientation of the site. Shelter behind hills from prevailing wind directions, and
- Create wind breaks by planting trees or making strong bush fences. Settlements with many trees experience lower wind speeds.

The layout of the buildings on the site can also influence the way winds affect them. Generally, settlements that are built in close clusters are known to suffer more damage than those that have a reasonable spacing between buildings. Large buildings can be used to shelter smaller buildings.

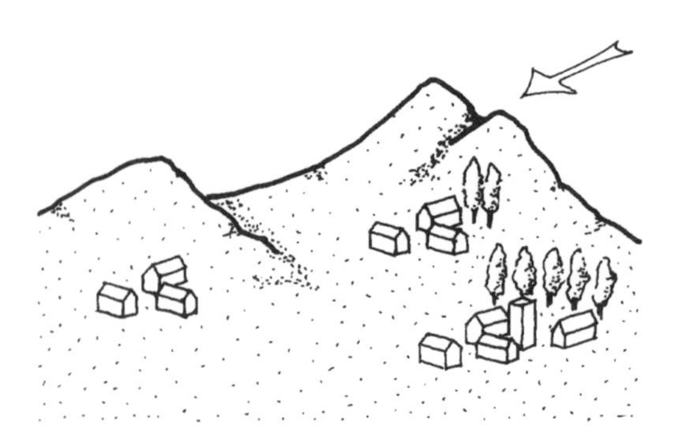

Figure 2.8 Sheltered sites help reduce the exposure to wind hazards.

A guide to safer layout would include:

- Site buildings some distance from adjacent structures (at least three times the plan dimension of the building)
- Site buildings in staggered formations rather than straight lines
- Keep buildings away from tall trees that might fall down, and
- Maximize street widths; where possible, they should be wider than 6m.

In cyclone-prone areas it is important to site to minimize the danger of flooding, both from the heavy rains that they bring with them and from the sea level rises they cause. This is discussed in the previous section on siting to minimize flood hazard.

The heavy rains that come with cyclones also trigger landslides and rockfalls. Siting measures to minimize the risk from potential instabilities of sites, as described in the previous section, are important in cyclone regions.

Safe siting in earthquake areas

Build on rock or stiff soil

Earthquakes, when they occur, affect a large area so it is difficult to find a site that will be unaffected by the seismic activity of the region. The most important element of safe siting in earthquake areas is to avoid being affected by land instabilities, covered in the siting section above, as earthquakes make these land instabilities a lot more likely to happen.

Firmer soil

Different types of ground do shake with different severities in an earthquake. Softer soils and those with high water content, generally shake more strongly than rocky sites. Where it is possible to site on firmer ground, this will reduce the severity of the vibrations experienced in an earthquake.

Liquefaction

One particular phenomenon that earthquakes sometimes cause, which can be damaging to buildings, is the liquefaction of soils. Loose soils on flat land, usually with a high water content, can suddenly lose their strength under strong vibrations in earthquakes. Soils effectively turn temporarily to liquid allowing structures that are built on them to sink or overturn. If your area contains sites with liquefaction potential it is best that further studies are carried out before large buildings or settlements are built on them.

Faults

In some areas earthquakes occur on a well-known *fault*, that is an identified geological break in the landscape where the earth ruptures take place. Where these are known, building across the fault or very close to them will almost certainly result in the destruction of the building if the fault moves. Capital-intensive infrastructure, hazardous facilities and other important buildings, should not be located in the vicinity of a known fault. However, the number of known faults that can be identified on the surface of the earth is very small. Most faults occur underneath the earth or rupture the ground somewhere over a wide area. The chances of a structure being affected by an unknown surface rupture in an earthquake are slight.

Further reading on safe siting

Coburn A.W. and Spence, R.J.S. (1992). *Earthquake Protection*, John Wiley, Chapter 6.

Jaffe, M., Butler, J., and Thurow, C. (1981). *Reducing Earthquake Risks: a planner's guide*, American Planning Association, Planning Advisory Service, Report Number 364, 1313 E. 60th St., Chicago, IL 60637, USA.

UNDRO, 1976, *Guidelines for Disaster Prevention, Volume 1. Pre-disaster Physical Planning for Human Settlements*, Office of the United Nations Disaster Relief Co-ordinator,

Wolfe, M.R., Bolton, P.A., Heikkala, S.G., Greene, M.M., and May, P.J. (1986). *Land Use Planning for Earthquake Hazard Mitigation: a handbook for planners*, Special Publication 14, Natural Hazards and Application Information Center, Institute of Behavioural Science, No. 3, Campus Box 482, University of Colorado, Boulder, CO 80309-0482, USA.

Chapter 3 Building Safely in Brick and Block

Masonry and hazards

Buildings made of masonry units, such as brick and concrete block, predominate in many parts of the world. Unit masonry is popular for urban low-cost housing because the material enables sound, durable buildings to be built at relatively low cost and using unsophisticated technology. Bricks and blocks and the skills of masons are found all over the world. In the past, solid burnt clay bricks have been the most common masonry units. However, because of costs of fuel and in some places limited supplies of suitable clays, they are being replaced by other types of masonry — hollow clay blocks, concrete blocks, stabilized soil and in some places dressed stone.

Masonry structures are, on the whole, resilient to damage from strong winds, if attention is paid to roof construction (see Chapter 7). A well-built masonry structure with good foundations is likely to survive moderately severe floods better than lightweight buildings, timber or earthen structures.

Masonry structures are, however, brittle and vulnerable to ground movement and subsidence. Small movements in the ground can cause structural cracks in the walls greatly weakening the structure, unless the foundations are strong enough to resist them (see Principle 2). Of all the hazards, however, masonry buildings are most vulnerable to earthquakes, especially when poorly built.

Earthquakes and brick buildings

Performance of brick and block materials in earthquakes has been mixed: the best built

Unreinforced brick masonry buildings are particularly vulnerable to earthquakes. House in Iran.

buildings have performed well, and sometimes better than reinforced concrete buildings alongside. But poorly built brick buildings have also contributed to some terrible earthquake disasters such as that of Tangshan, China, in 1976. Multi-storey brick masonry buildings also have a particularly poor record of damage in Californian earthquakes.

As a result of the lessons learnt from these earthquakes, some principles for sound building in earthquake areas have become established, and these principles are the subject of this chapter. In brief they can be stated as follows:

1. Robust building form
2. Firm foundations
3. Good quality materials
4. Strong walls
5. Distributed openings
6. Horizontal reinforcement
7. Safe modifications, and
8. Regular maintenance.

Each of these principles is elaborated in this chapter.

PRINCIPLE 1

Robust building form

Lay out the plan of the building so that it is regular and symmetrical, with closely spaced walls running in both directions.

Experience shows that the overall shape and internal planning of a building are important factors influencing its resistance to the forces of earthquakes and high winds. Buildings which are substantially unsymmetrical or irregular in plan are much more likely to be seriously damaged in earthquakes than regular buildings. Also, the resistance of masonry buildings is greatly enhanced if the walls are of stocky proportions, closely spaced and arranged to give each other mutual support.

Earthquakes shake buildings in all directions — upwards and downwards, backwards and forwards, and from side to side. The earthquake shaking sets up internal forces in a building which cause its elements to distort – to twist, stretch and bend – in ways which are quite different in each earthquake and are, therefore, difficult to predict. The principal structural elements of buildings which are irregular or unsymmetrical are much more likely to be damaged in earthquakes than those which are symmetrical and regular.

Experience shows that *small* buildings are much more likely to suffer damage if they:

- Have irregular, L-shaped or T-shaped plans
- Have the openings in their external walls concentrated either on one side or on two adjacent sides of the building, and
- Have an upper floor which overhangs the lower floor.

The planning of the walls and their layout is just as important as getting the right overall shape for the building. Masonry walls are much better able to resist shaking along their length than shaking which occurs at right – angles to their length. Walls which are arranged to give each other mutual support by being connected to perpendicular walls at each end and at regular intervals along their length are less likely to fail under earthquake shaking than long lengths of unsupported wall. Also, high walls and thin walls are in general more vulnerable than low walls and thick walls. To have the best chance of resisting earthquake ground shaking, the walls should be thick in relation to their height.

General principles:

- Make the building as symmetrical and regular in plan as possible
- The overall plan should be roughly rectangular; it should also be compact, ie not long and thin
- Arrange the walls to give each other mutual support
- Each wall should be *either* connected to a crosswall at regular (5 to 7m maximum depending on wall thickness) intervals *or* supported by piers or buttresses *or* cranked in plan
- For two-storey buildings use the same wall layout on each of the two floors, and make the amount of wall opening on each floor about the same
- Divide up T-shaped and L-shaped plans into separate, structurally independent pieces
- Buildings or structurally independent parts of buildings should be separated with an open gap of at least 75mm, and
- The area of opening in each of the walls should be kept to a minimum and equal in opposite walls.

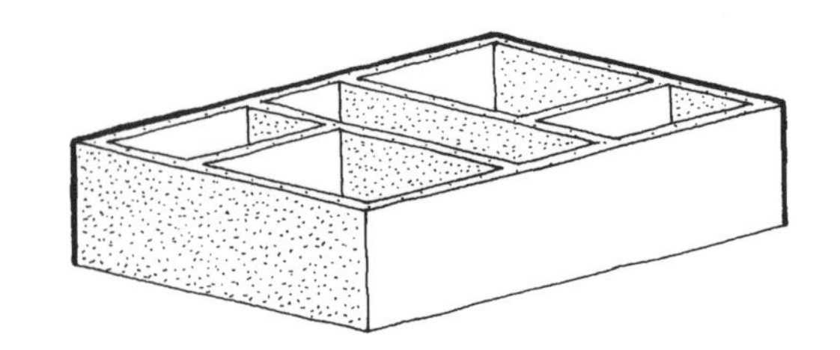

Figure 3.1 Make building plan regular and compact. Each wall should be connected to a cross wall at regular intervals.

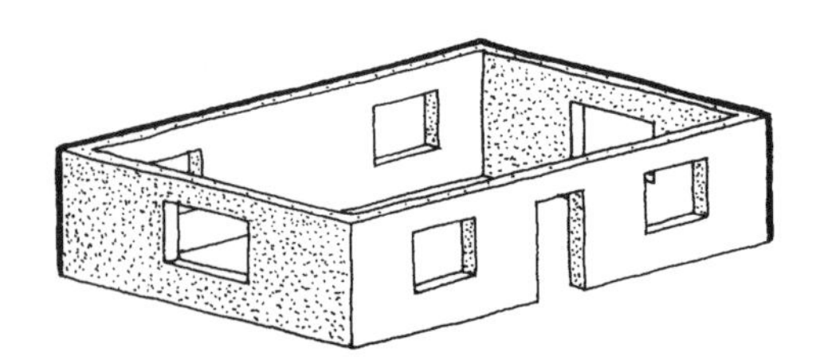

Figure 3.2 Keep openings small and equal in each wall.

PRINCIPLE 2

Firm foundations

Make sure that the building has foundations which are well-connected and can spread the loads from the building into firm ground.

Well-built foundations form the basis of safe construction. Continuous strip foundation of rocks laid in cement mortar. Government construction, Pakistan.

Well-built foundations form the basis of safe construction. Strong foundations bind the building together, provide solidity against lateral pressures and may prevent the structure from being damaged by ground movement.

Good foundations should provide a measure of resilience against the pressure of flood waters and survive the erosion of sub-soils. Subsidence or weak ground can cause damage and weakening of the structure of a building unless the foundations are adequate to withstand it. They should also be built to protect the walls from deterioration through seepage of ground water or other agents.

Most importantly for brick buildings, well-built foundations can protect a building from the worst effects of earthquake ground motion through the underlying soil.

The foundations of a building bind the bases of the walls together and distribute the loads from the strong materials used for the building into the softer and variable soils beneath. They need to be able to transmit the loads from the sideways movement of the building into the ground and to withstand distortions of the ground without destroying the walls above.

Buildings on foundations which are too narrow, discontinuous or are based on poor or variable soil are liable to be severely damaged by earthquakes however well-constructed the walls are.

Foundations are likely to be in contact with ground water which will weaken them over time if they are not built in durable materials. In some areas they will be subject to frost.

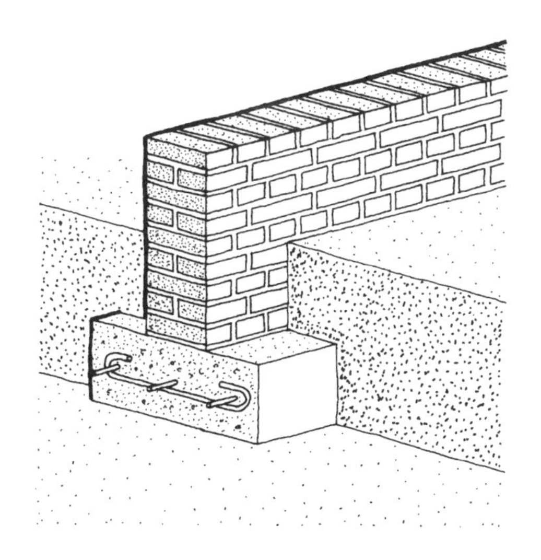

Figure 3.3 On soft soil, use reinforced footings shown here or a plinth band (Figure 3.4).

General principles:

- If your site is on *firm soil,* use continuous strip footings
- If your site is on *soft soil* reinforce the foundations or provide a reinforced *plinth band*
- Make the foundations of well-compacted concrete with a minimum cement content ten per cent by weight of the aggregate
- If the ground is subject to frost, build the base of the foundations at a level below the freezing zone, and

- Protect the foundations and subsoil from becoming saturated with water.

PRINCIPLE 3

Good-quality materials

Build in the best quality materials available.

Walls form the structure of a brick building. The quality of the walls is only as good as the materials they are made from. Good materials make a building safer—safer against wind loads, safer from floods, safer against accidental impacts or unusual surcharges of load. Walls can suffer damage during wind storms from the effect of direct pressure, or from suction forces that act out-of-plane, pushing the wall in its weakest direction.

Under earthquakes, the quality of the masonry materials becomes critical: if the masonry materials are weak, disintegration of the walls can occur quickly after the onset of a strong earthquake. Occupants of the building may have little time to escape, and may be trapped under falling masonry.

Buildings built from weak masonry units, such as poor-quality bricks or badly made concrete blocks are extremely vulnerable to earthquakes, even when other principles of building safely have been observed. Equally, even if

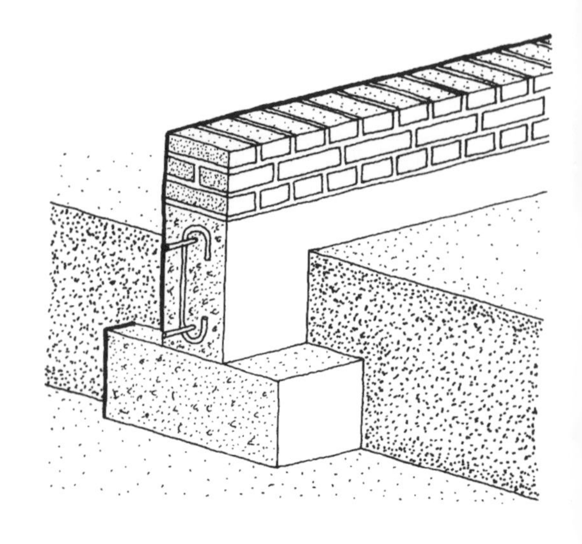

Figure 3.4 On soft soil use re-enforced footings or a plinth base.

Use the best quality materials available. Hollow concrete block construction. Turkey.

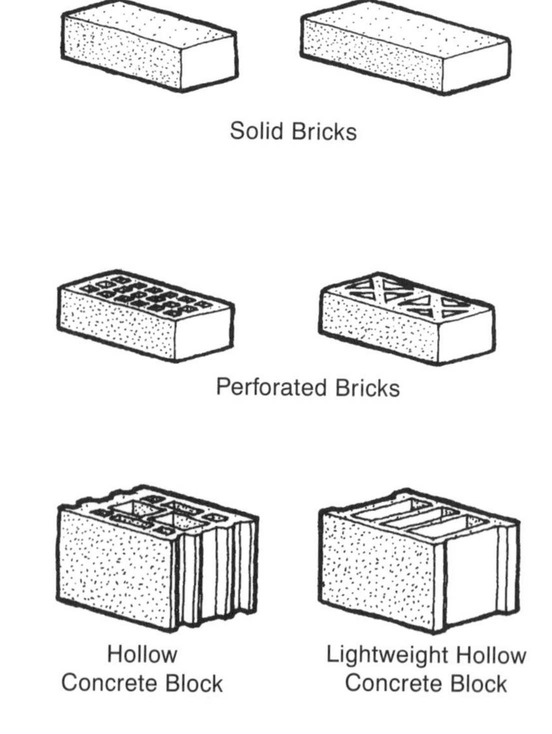

Figure 3.5 Good-quality bricks or blocks are important in building robust masonry.

the masonry units are sound, if they are put together using a very weak mortar, earthquake forces can cause the wall to disintegrate.

The strength of masonry materials can also deteriorate with time as a result of the action of rain, frost or ground water, and walls built with them become increasingly vulnerable.

General principles:

Choosing materials

A great variety of masonry materials are suitable for use in earthquake areas. These include:

- Fired clay bricks
- Concrete blocks
- Stabilized soil blocks, and
- Dressed stone blocks.

Earthen and rubble stone masonry may also be used with special precautions; these are discussed in Chapter 4.

Masonry materials:

- Fired clay bricks or concrete blocks may be solid or hollow. Hollow blocks are good because they can be vertically reinforced
- Whenever possible, masonry materials used should be made to satisfy national standards. These always include requirements for minimum strength, and often requirements for shape, appearance and durability, and
- Look for bricks which make a ringing sound when you strike them together, do not break when dropped, and do not disintegrate when soaked in water.

Mortar materials:

- The quality of the mortar is of primary importance. Mortars should be made of sand, lime or cement and water. Mud mortars should be avoided. Mortars are used to bind the masonry blocks and prevent them from sliding or disintegrating during the horizontal earthquake shaking. Therefore, it is very important that the mortar mix is strong enough to counter those forces
- The mortar mix used should have a minimum ratio of 1:2:9 cement:lime:sand, or 1:6 cement:sand if lime is not used
- Sand should be clean, free from clay and plant matter. River sands are better than beach sands. Water should be clean
- Lime and cement should satisfy national standards for quality, and
- If weak masonry units are used, a strong mortar will *not* improve the strength of the brickwork as a whole, but *will* make it more liable to crack. This can be particularly dangerous in earthquake areas.

PRINCIPLE 4

Strong walls

Walls should be well-bonded, not too thin.

The resistance of the vertical structure has a major influence on a building's structural integrity, and in the case of masonry buildings this is provided by the strength of the walls. In addition to the strength of the masonry units themselves and the strength of the mortar, the strength of a masonry wall as a whole is influenced by the way the masonry units are laid or bonded and by the size and location of the openings in it. The skill of the builder is crucial in this case. Training to learn stronger bonding techniques is always to the benefit of builders and the community.

Masonry walls often fail or are severely damaged in earthquakes because, even though they are made of good materials, they are poorly bonded. If the vertical joints are not staggered, loads cannot be properly distributed through the wall, and vertical cracks can easily develop.

If a masonry wall is a cavity wall consisting of two leaves of brickwork or blockwork tied together, the connection between the two leaves is critical. Outer leaves of brickwork can collapse because wall ties have been omitted, or have corroded over time, or are inadequately bonded into the brickwork.

Good bonding is especially important at the wall-to-wall junctions, since the mutual support which walls running at right angles to

Strong walls are thick and built with good bonding of the bricks. Additional strength can be obtained from adding steel reinforcement. Confined masonry in Pakistan.

each other can give depends very much on good bonding. This is particularly difficult to achieve if one wall is built to its full height before the adjoining wall is built or if the two connecting walls are built of different masonry materials.

Increasing levels of strength can be added to masonry walls:

- **Standard (unreinforced) masonry** relies on bonding of the masonry units alone for its strength and integrity
- **Masonry with ring beam** has horizontal reinforcement at the top of walls, at floor – levels or at other levels to tie walls together and give increased lateral resistance
- **Confined masonry** has vertical reinforcement at corners and at the sides of openings, increasing the compression of the masonry to provide greater strength, and
- **Reinforced masonry** has both vertical and horizontal reinforcement distributed throughout the wall, tying the masonry into a coherent structure.

Wherever possible, and if it can be afforded, confined or reinforced masonry will provide the strongest structure. Methods of designing and building confined and reinforced masonry are described in other publications, detailed in the further reading lists. For small buildings, the high cost of reinforcement and the skills needed may make confined or reinforced masonry unattainable. A minimum requirement for masonry in earthquake areas is the addition of a ring beam, described in Principle 6. With or without reinforcement, safe masonry requires good bonding techniques and care in construction.

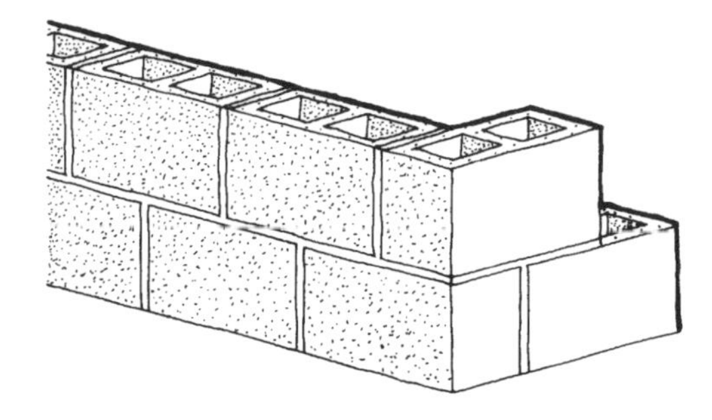

Figure. 3.6 If building loadbearing walls in a single leaf of masonry, ensure they are well bonded and at least 20cm thick.

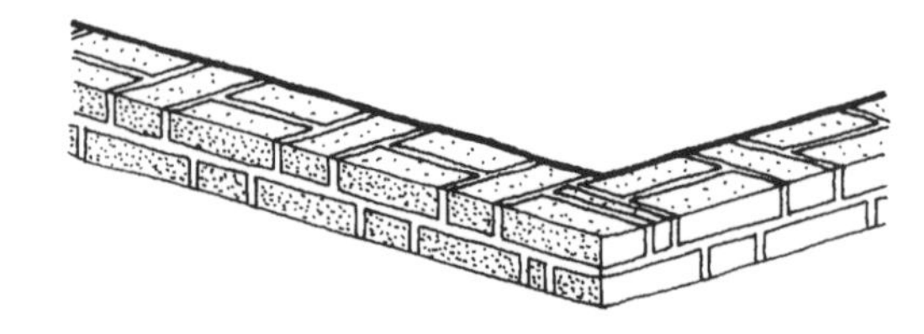

Figure. 3.7 A stronger wall can be built using a double leaf of masonry cross-bonded through its thickness.

General principles:

- Masonry walls may be made from a single leaf of masonry units (not less than 20cm thickness), or
- Two parallel widths bonded together.

Whichever system is used:

- The ratio of the height of the wall to its overall width should be limited to avoid slenderness, and
- The masonry units should be properly bonded.

Specific recommendations:

- Precise rules on slenderness and bonding depend on the type and width of masonry used
- All the walls in any storey should be built up simultaneously, building in stages of about 1m height
- Ensure full bonding around corners and between main walls and cross walls or partition wall, and
- For additional safety, strengthen corners by adding reinforcing bars or wire mesh (wire fabric) in the horizontal mortar joints.

PRINCIPLE 5

Distributed openings

Openings should be small, well-spaced and located away from corners.

Openings are required in all masonry walls, but they are also cuts in the structural fabric of the wall. There is plenty of reserve strength in most masonry, so openings do not cause a serious weakening as long as they are relatively small in size, and adequate widths of masonry wall are left between them.

Openings which are very close to the end of a wall, or to a cross wall, cut into the corner area where the wall needs stiffening, so openings should be kept away from corners. Where openings are too large or badly located, serious damage to the wall can begin at the openings and spread through the whole structure.

In wind hazard areas, large glazed areas should be avoided or divided into smaller sections. Generally try to limit the amount of window openings and provide shutters to protect the windows.

All openings should be provided with an adequate reinforced lintel, and where possible the sides should be reinforced.

General principles:

- Make sure that the size of openings in the main resisting walls is as small as possible
- Locate openings away from the corners
- Space openings in any one wall to leave masonry piers of adequate width between them
- For two storey buildings, locate the openings in the upper storey above those in the lower storey, and
- Provide each opening with a reinforced lintel.

Specific recommendations:

- Locate any opening a minimum of one quarter of its height (and not less than 60cm) away from the inside face of the wall at a corner
- The horizontal distance (masonry pier width) between two openings should not be less than half the height of the shorter opening, and not less than 60cm
- The total width of all openings should not

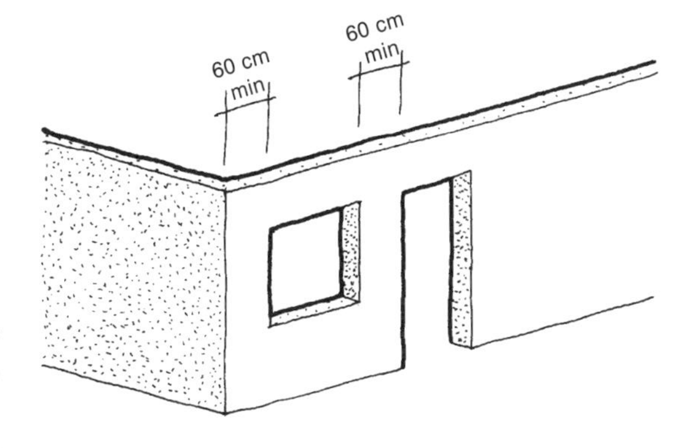

Figure 3.8 Keep openings away from each other and from corners.

exceed half of the distance between cross walls in any one length of wall

- For a two storey building, the total width of all openings should not exceed forty per cent of the distance between cross walls on any one length of wall
- The vertical distance from an opening to another directly above it should not be less than half the width of the smaller opening and not less than 60cm
- For additional safety, place reinforcing bars in the brickwork around each opening. (Methods for doing this depend on type of brickwork.)

PRINCIPLE 6

Horizontal reinforcement

Tie the tops of all walls and the floor or roof plane together with a ring beam.

The single most common cause of failure in masonry buildings is the separation of the walls at corners, because they are not adequately tied together. This then allows perpendicular walls to shake independently, and they fail by overturning.

Experience shows that the safety of masonry buildings in earthquakes and high winds is greatly improved if the tops of all the walls are connected by a continuous ring beam or tie beam.

This has the additional benefit of stiffening the tops of the walls, protecting them from cracking, and enabling lateral forces to be transmitted into all the available walls through the floor or roof. But if the ring beam is not securely fastened to the wall and the floor or roof adjacent, the walls can separate from the beam under severe shaking.

General principles:

- Provide a continuous ring beam of reinforced concrete to the tops of all walls
- Make sure that the ring beam is properly tied to the floor or roof which is supported on it or at the same level
- Make sure that the ring beam is stiffened at the corners, and

A concrete ring beam holds a masonry building together that would otherwise have collapsed. Earthquake, Turkey.

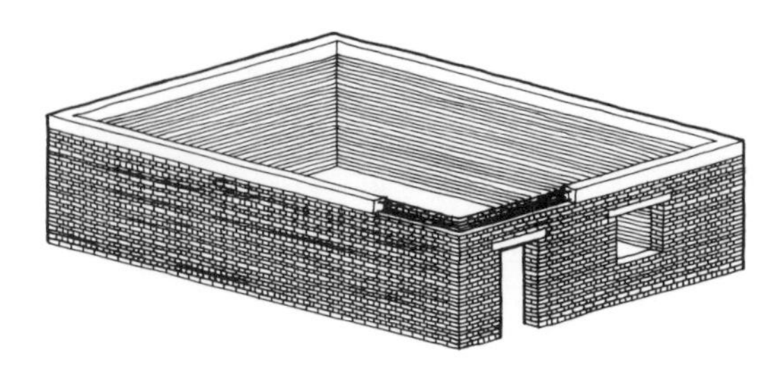

Figure 3.9 A reinforced concrete ring beam greatly increases the safety of masonry buildings in earthquakes and high winds.

- For additional safety in walls more than 2.5m high use an additional ring beam at the level of the lintel.

Construction of a concrete ring beam

- A concrete ring beam should be made of well-compacted concrete of mix 1:2:4, cast

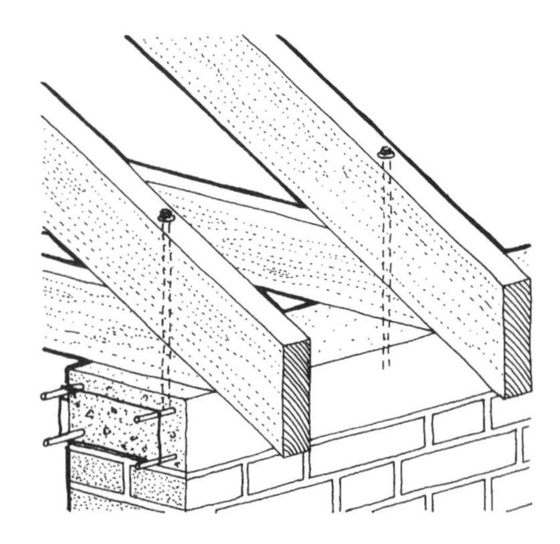

Figure 3.10 Anchor the roof to the concrete ring beam.

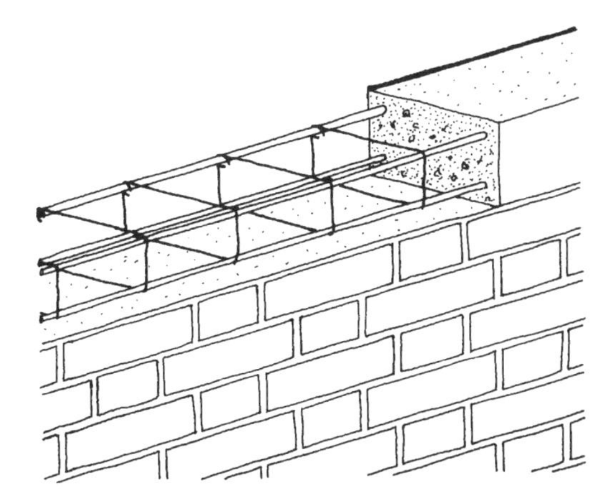

Figure 3.11 Make sure that the reinforcement in the ring beam is continuous, well-spaced and with regular hoops.

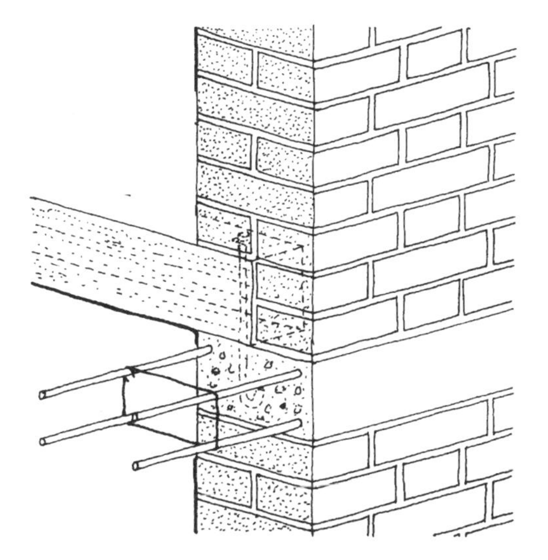

Figure 3.12 Anchor the floor joists into the ring beam.

into timber formwork. Its width should be the width of the wall (not less than 20cm), and its thickness a minimum of 75mm

- Reinforcement should consist of at least two 10mm diameter steel bars located close to each face of the wall with a minimum 25mm cover to each face
- For buildings where the distance between cross walls is greater than 5m, or the earthquake risk is particularly high, larger amounts of reinforcement should be used

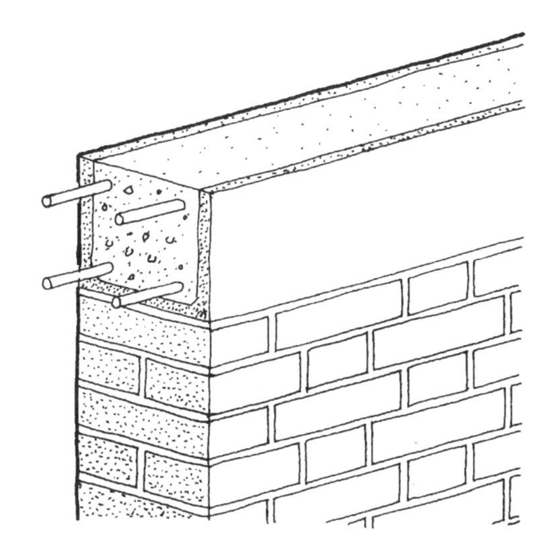

Figure 3.13 An alternative ring beam can be cast inside hollow concrete blocks.

- At corners the reinforcing bars should be arranged to give full continuity
- Floor joists or roof trusses should be bolted down to the ring beam using cast-in bolts
- If steel reinforcing bars are unavailable, a timber ring beam may be used instead. (for construction of a timber ring beam see Chapter 4, Principle 3), and
- Where hollow block masonry is used, the ring beam may be cast inside U-shaped concrete blocks.

PRINCIPLE 7

Safe modifications

When altering or adding to a building, do so in a way which strengthens rather than weakens it.

Where old buildings have been modernized and enlarged for new uses, this is often done by removing walls, changing the location of openings and adding extra storeys. All these actions change the structural resistance of a building.

Buildings are frequently modified by introducing new services, such as water, electricity, drainage, which usually involves cutting into the structure and often weakening it.

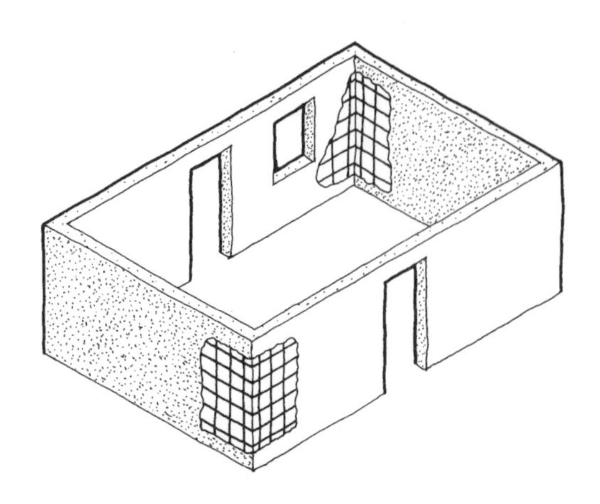

Figure 3.14 Steel mesh applied outside and inside, and rendered with cement, can restore strength to a cracked wall.

Cutting an old wall introduces new cracks into it which can be easily exploited and extended by an earthquake or other surcharge causing the whole wall to fail.

On the other hand, the modification of existing buildings offers an opportunity to introduce strengthening measures to bring the structural resistance up to acceptable level. If this is done at a time when other modifications are being carried out, the extra cost may be comparatively low.

In addition, when modifications or extensions are being planned the desirability of structural symmetry must always be kept in mind. Alterations that make a building less symmetrical in either a horizontal or vertical direction will not be to its advantage during an earthquake, and are likely to deteriorate its performance.

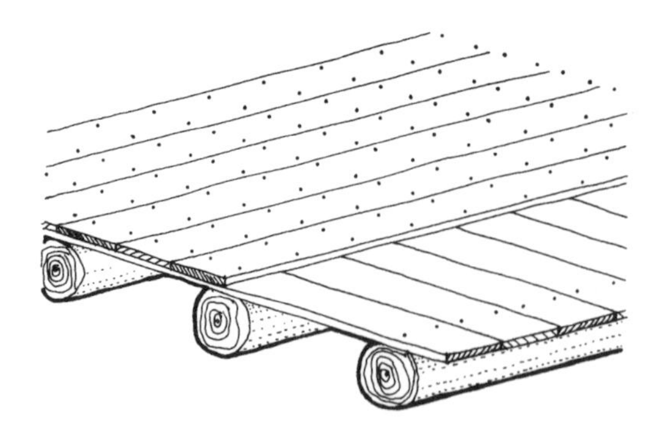

Figure 3.15 Floors can be stiffened by adding an extra layer of concrete or floorboards.

General principles:

When altering an existing building, use the chance to improve its overall structural resilience. This may involve any or all of the following changes:

- Improving the arrangement of walls to improve its regularity or symmetry
- Strengthening of existing walls by filling cracks or adding external reinforcement
- Improving the bonding between walls by adding ties
- Adding a ring beam to connect the tops of the walls, or strengthening an existing ring beam
- Tying floors or roofs to main walls, or ring beam
- Strengthening existing floors by the addition of an extra layer of concrete or floorboards
- Improving the arrangement in distribution of openings in walls, and
- Strengthening foundations.

Always:

- Remove existing masonry with minimum cutting and reinforce around new openings
- Bond new walls to existing walls using new steel ties.
- Tie new floors or joists to existing walls, and
- Ensure that extensions which alter the plan form are structurally independent.

PRINCIPLE 8

Regular maintenance

Regularly inspect, protect and repair each of the structural elements of the building.

The record of past damage by earthquakes shows that in any area hit by an earthquake, the worst-damaged buildings, and the highest concentration of casualties, occur in buildings which are poorly maintained and have been allowed to deteriorate.

All buildings deteriorate unless they receive maintenance, and in time, lack of maintainance has a serious effect on a building's ability to resist an earthquake.

In the case of masonry buildings the ways in which poor maintenance can reduce a building's earthquake resistance include:

- Weakening or erosion of masonry units or mortars (particularly at the base of walls) due to effects of water or frost, or of plant growth
- Corrosion of wall ties or roof fixing bolts
- Corrosion of reinforcing bars in ring beams due to cracking of concrete or inadequate cover to steel
- Rot or fungal growth on roof or floor timbers
- Cracking of walls due to foundation movement, and
- Damage to any element of the structure due to the effects of previous small earthquakes.

Regular inspection of a building by the owner and maintenance where needed is a crucial aspect of keeping a strong building, resistant to the various hazards that might affect it in its lifetime.

General principles:

Inspect structure:
Inspect all parts of the building regularly giving particular attention to:

- Possible deterioration of mortar near the ground, near downpipes, in chimneys
- Deterioration of masonry units themselves
- Possible vertical cracks in corners, especially under the cornice
- Corrosion of any visible fixings of roof or floor joists, windows, or any of these members loosening
- Rotting of timbers, especially close to ground or under roofs where they may be damp, and
- Cracking of concrete or exposure of reinforcing bars.

Inspect services:

- Inspect (or have suppliers inspect) all electricity , gas or liquid fuel fittings, and
- Inspect external drains.

Identify problems:

- Identify probable causes of the deterioration: leaking roofs, lack of site drainage, leaking gutters, poor concrete, foundation settlement are frequent examples.

Repair and protect:

- First, eliminate the cause of the damage (leaking, drainage) if possible, and
- Then, where needed, repair or replace the damaged elements. Make sure you use strong materials.

Post-earthquake:

- In the event of an earthquake tremor inspect the building for structural damage, and repair non-superficial damage. Call for specialist help when damage is serious.

Summary

Make sure that:

- The building is symmetrical in plan
- Walls are arranged to provide mutual support
- Foundations are well-connected together
- Bricks and mortar are of the best quality available
- The builder is skilled and dedicated
- Openings are small and well-spaced, and
- The ring beam is well-tied to the walls and roof.

Avoid:

- Building on very soft ground or on fill
- T- and L-shaped building plans
- Masonry units which are cracked or crumbly
- Using earth or mud mortars
- Mixing different masonry materials in one building
- Large openings in major supporting walls
- Concentrating openings in one or two adjacent sides, and
- Any discontinuity in the ring beam.

Further reading on building safely in brick and block

Daldy, A.F (1972). *Small Buildings in Earthquake Areas.* Building Research Establishment, Garston, Watford WD2 7JR, UK.

Dancy, H.K., (1980), *A Manual of Building Construction*, IT Publications.

International Association for Earthquake Engineering (1986). *Guidelines for Earthquake-resistant Non-engineered Construction*, IAEE, Kenchiku Kaikan, 3rd Floor, 5-26-20 Shiba, Minato-ku, Tokyo. Chapter 4.

Spence, R.J.S., and Cook, D.J. (1983). *Building Materials in Developing Countries*, JohnWiley.

UNDP / UNIDO (1984). *Building Construction under Seismic Conditions in the Balkan Region*, Vol. 3 Design and Construction of Stone and Brick Masonry Buildings, UNIDO, Vienna.

Windass, M. (1981). 'ARTIC and the Construction of Houses in Andhra Pradesh Following the Cyclone and Sea Surge of November 1977', in Davis I. (ed.), *Disasters and the Small Dwelling*, Pergamon, Oxford.

Chapter 4 Building Safely in Earth and Stone

Weak masonry and hazards

Sun-dried earth blocks, rammed earth, stone and rubble have historically been some of the most widely used and versatile building materials. They are still widely used today.

Low-strength masonry using various forms of earthen construction or rubble stone walls is very common in some of the world's most hazard-prone regions, traditionally across much of southern Europe, the Middle East and Indian sub-continent. It is still widely used today in many parts of Latin America, Asia and elsewhere across the world.

The mass of low-strength masonry makes it comfortable in hot arid climates. The cheap availability of the raw materials means that they will continue to be used in the housing of the lower-income groups, and the inhabitants of rural or less-industrialized areas.

However, these types of structures are extremely vulnerable to weathering, erosion from rain, subsidence and other factors.

Exposure to water in floods, both from saturation and from erosive flows, is also very destructive because the earthen blocks and earth mortars are easily dissolved and eroded. Because of their weight they are less vulnerable to wind damage, but roofs (see Chapter 7) may easily be damaged if not firmly anchored to the building.

Earth and stone masonry buildings are also notoriously vulnerable to earthquakes: con-

Earth and stone buildings are notoriously vulnerable to earthquakes. They collapse easily and have caused high numbers of casualties. Rubble masonry collapse in an earthquake, Iran.

siderable damage and loss of life has occurred in areas where these materials were used. The main causes of failure are: disintegration and separation of walls; overturning of walls; separation of corners of walls; and separation of roofs from supporting walls. The collapse of the heavy flat roof resulting from these failures leads to burying and killing the occupants.

The large death tolls in past disasters have given these materials a poor reputation. It is certainly more difficult to achieve good earthquake resistance with these materials than with other forms of masonry because of the variability and low strength of the masonry. But as long as special precautions are taken these materials can be used to build houses and other buildings which have a satisfactory earthquake resistance, to all but the most extreme earthquakes.

The different forms of earthen construction include rammed-earth, adobe block, wet mud construction and plastered frame construction. Local traditions and the soil types used vary widely. There is also a wide variety of types and qualities of stone construction in use. Because the local traditions vary so widely it is not possible to give adequate guidance in a single document. But a few essential principles apply to all low strength masonry construction. These may be summarized as:

1. Good practice
2. Robust layout, and
3. Ring beam.

Some of the more general guidelines for achieving earthquake resistance are summarized below. Using these principles will also protect them from the other major hazards. But for more specific guidance, see the list of further reading.

Figure 4.1 Lay stones in level courses between squared corner stones. Use a plumb line to check walls are vertical.

Figure 4.2 Use long 'through-stones' that lie across both leaves of stone, or dog-tooth stones that inter-link from one leaf to another.

PRINCIPLE 1

Good practice

Follow good construction practice.

The most effective way to ensure good protection from earthquake and other hazards is to follow good construction practice. In any community struck by a disaster, the well-built houses have been found to have a much better chance of survival than those of exactly the same size and shape which are poorly built.

Because of the huge variety of local traditions, good practice for earthen and stone masonry cannot be as easily summarized as for brick and concrete construction. Three of the most common forms of construction are:

- Rubble stone masonry
- Adobe blocks construction, and
- Rammed earth construction.

Some of the principal considerations for each follow.

Rubble stone masonry:

- Use a stable mortar of earth which has low shrinkage or (preferably) earth-lime or earth-cement. Fill all joints
- Lay all stones flat, in level courses, and in good contact with at least two other stones
- Key all internal walls into external walls with longer bonding stones
- Use long so-called 'through-stones' passing through the full width of the masonry at regular intervals to tie the inner and outer leaves together
- Stagger vertical joints between stones so that a large vertical crack cannot occur.
- Use large dressed (rectangular) stones at all external corners, and where possible, around openings

Select soil for building with great care. Avoid shrinkable clays and seive soils before use. Soil sifting in Turkey.

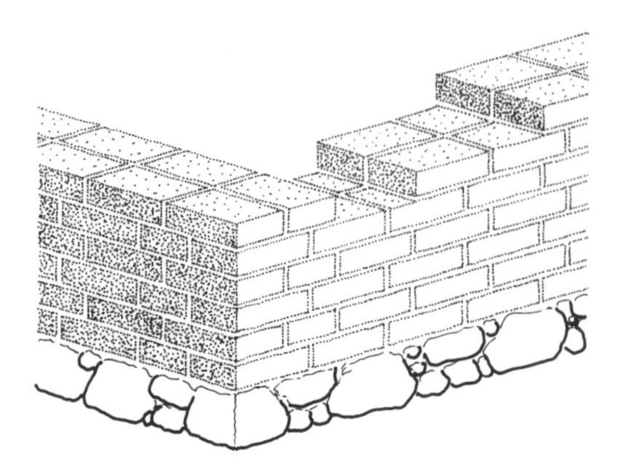

Figure 4.3 Adobe block construction should be massive, well-bonded and with protective courses at the base.

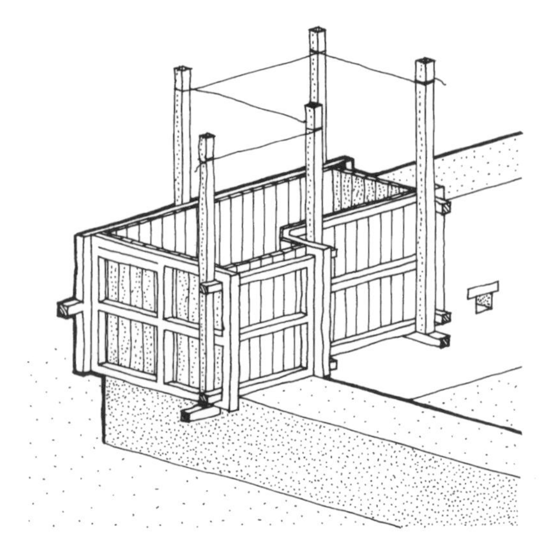

Figure 4.4 Corners are the weakest part of rammed earth construction — build corners in a single L- or T-shaped mould.

- Prevent water penetration into the wall from the roof, and
- Protect the base of the wall from water by surface drainage.

Adobe can be used to make a stable wall if it is thick, laid in horizontal courses and well-bonded. Protective courses of stone at the base and a horizontal ring beam give additional safety. Undamaged house in an earthquake in Iran.

Adobe block construction:

- Select a soil which combines good dry strength with low shrinkage; use established field and laboratory tests to identify a suitable soil; blend soils from different sources if needed
- Use the same soil for mortar as that used for the block construction
- Add straw or other fibres to the soil to improve strength and control shrinkage
- Build in level courses
- Stagger the vertical joints between consecutive courses
- Fill all horizontal and vertical joints with mortar
- Build up external and internal walls together; bond all walls at corners and intersections
- For additional strength and durability use bitumen-stabilized adobe blocks, and
- Protect the base and the top of the wall from moisture penetration. Where possible, build adobe walls on a course of stone or brick at ground level to avoid ground water weakening the wall.

Rammed earth construction:

- Select a soil which has low shrinkage combined with good compacted density
- Use established field and laboratory tests to check soil suitability; blend soils from different sources if needed
- Find the correct water content for compaction by testing and keep close control of it
- Use *through ties* to ensure that the formwork is strong enough to withstand compaction pressures without bowing
- Use formwork of maximum 1 to 1.5m height, with T- or L-shaped moulds for corners
- Stagger vertical joints in adjacent lifts to prevent formation of continuous vertical cracks
- For corner strengthening, build moulds with an internal 45° chamber: this can be repeated on the outside, and
- Protect the base and the top of the wall from moisture penetration.

PRINCIPLE 2

Robust layout

Arrange the building to provide a compact unit.

The following recommendations are specific to earthquake protection.

General recommendations:

- Build houses of one storey only
- Use an insulated lightweight roof in preference to a heavy compacted earth roof
- Arrange the wall layout to provide mutual support by means of cross walls and intersecting walls at regular intervals in both directions, or use buttresses
- Keep the openings in the walls small and well-spaced, and
- Build on firm foundations as described in Chapter 3, Principle 2.

Specific recommendations:

- The wall height should not exceed eight times the wall thickness at its base, and in any case not be greater than 3.5m, and

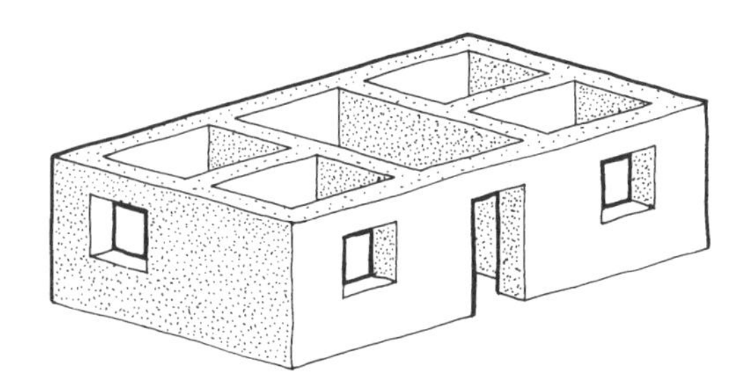

Figure 4.5 The safest building form is a squat, single storey house, with small windows and a regular, compact plan with frequent cross-walls. Build a lightweight roof if possible.

- The unsupported length of wall between cross walls should not exceed 10 times the wall thickness, with a maximum of 7m.

Openings:

- Wall openings should not exceed one-third of the total wall length
- No opening should be wider than 1.2m
- Bearing of lintels on walls should be at least 500mm, and
- Provide piers of at least 1.2m width between openings.

A timber beam adds considerable strength to a stone or earth structure making it safer in earthquakes. Building strengthening project, Turkey.

PRINCIPLE 3

Ring beam

Provide a ring beam to tie together the tops of all walls.

The following recommendations are specific to earthquake protection.

A ring beam which ties together the tops of the walls is the most essential component of the earthquake resistance. It must be strong, continuous and well-tied to the walls and the roof. The ring beam can either be a concrete ring beam or a timber ring beam. The construction of a concrete ring beam is described in Chapter 3, Principle 6. The construction of a timber ring beam is described below.

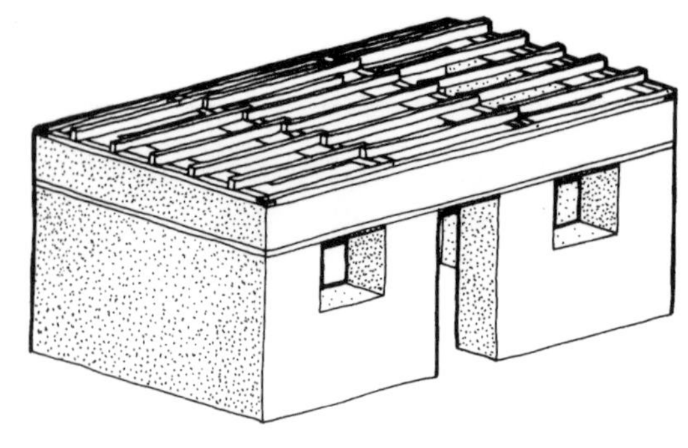

Figure 4.7 The timber ring beam is most effective placed at lintel-level. The tops of the walls should support a timber wall-plate, firmly fixed to each roof beam.

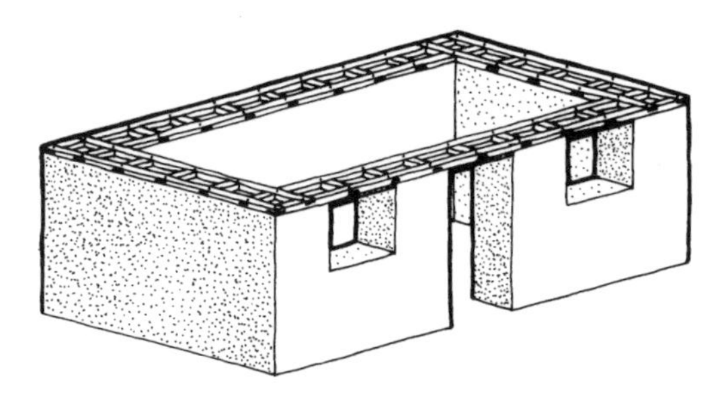

Figure 4.6 A timber ring beam should be built from two timbers connected by cross members. All joints should be lap-jointed and nailed. The timber roof beam is most effective at lintel level.

General recommendations:

- Provide a continuous concrete or timber ring beam to tie together the tops of all walls
- Ensure the continuity of the ring beam by lapping the reinforcement or splicing the timber
- Make sure that each roof member is separately connected to the ring beam which supports it at each end
- The ring beam should be strengthened at all corners, and
- For walls greater than 2.5m in height, an additional ring beam should be provided at lintel-level.

Construction of a timber ring beam

(For construction of a concrete ring beam see Chapter 3, Principle 6):

- For adobe construction in areas of moderate seismicity, the ring beam can consist of a single timber member of minimum dimensions 50 x 100mm set at the middle or outside face of the wall
- For all random rubble stone, and adobe or rammed earth in areas of moderate or high seismicity, the ring beam should consist of a pair of parallel timbers, 50 x 100mm, tied together at intervals with cross timber members connected by nailing
- Corners should be strengthened by additional nailing or diagonally braced with a timber of the same size attached to the main timbers with halved joints
- The best location in the wall for a timber ringbeam is at window lintel-level. The masonry above the lintel should be secured by a timber wall-plate fixed to every roof beam
- Tie each separate roof member to the ring beam at each end, and
- Where lightweight roofs are used (sheet or fibre concrete tiles), the ring beam should be tied into the walls at regular intervals.

Summary

Make sure that:

- The walls are as low as possible
- Walls are arranged to give each other mutual support
- Foundations are in water-resistant materials
- Materials are carefully selected
- Established good construction practice is followed
- Openings in the walls are small and well-spaced, and
- The tops of all the walls are tied together with a continuous ring beam with good tensile strength.

Avoid:

- T- and L-shaped building plan shapes
- Straight vertical joints, especially at corners
- Mortars which have a high shrinkage
- Unbonded walls

Further reading on building safely in earth and stone

Daldy, A.F (1972). *Small Buildings in Earthquake Areas.* Building Research Establishment, Garston, Watford WD2 7JR, UK.

Spence, R.J.S. (1971). *Making Soil Cement Blocks,* Commission for Technical Education and Vocational Training, Zambia.

International Association for Earthquake Engineering (1986). *Guidelines for Earthquake-resistant Non-engineered Construction,* IAEE, Kenchiku Kaikan, 3rd Floor, 5-26-20 Shiba, Minato-ku, Tokyo. Chapters 7 and 5.

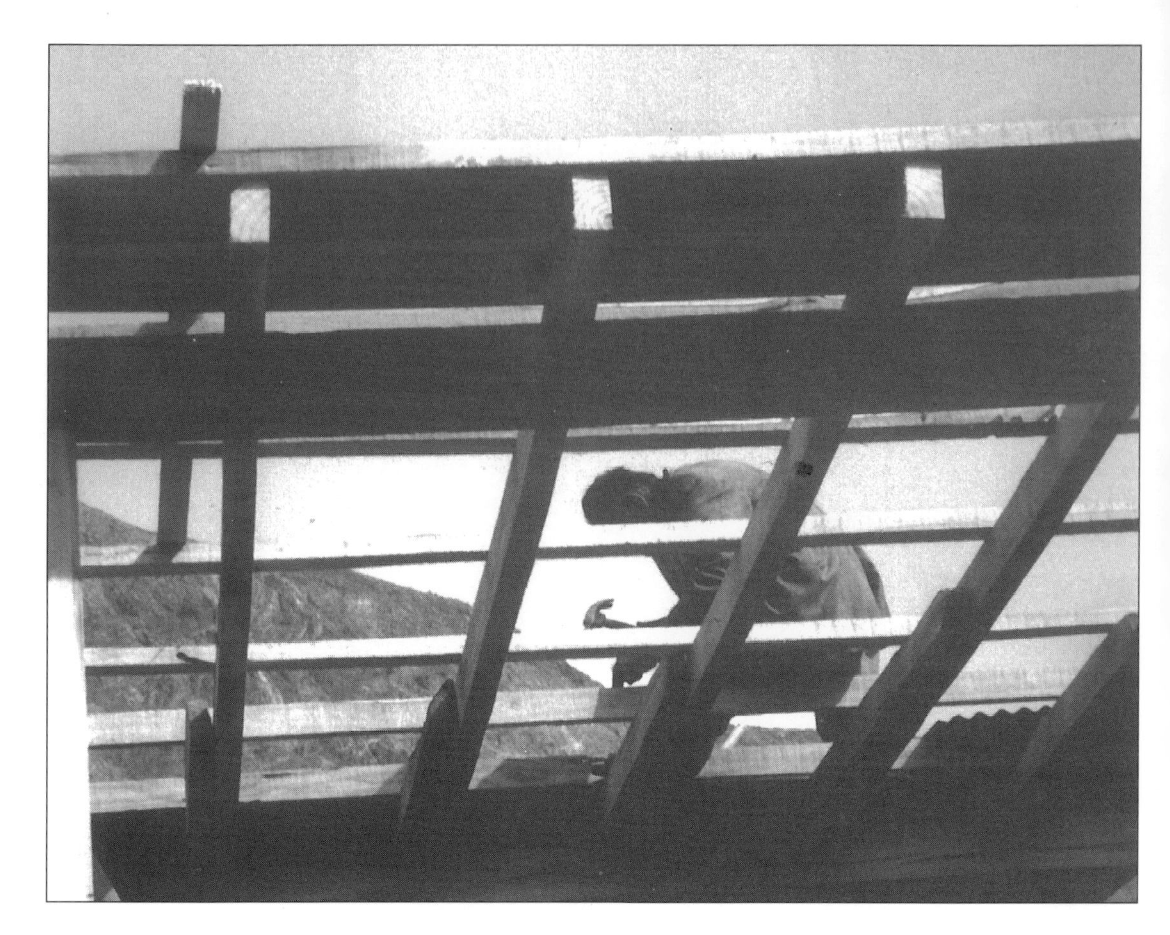

Chapter 5 Building Safely in Timber

Timber and hazards

Timber and bamboo are traditional building materials common throughout the world and excellent for low-cost housing construction. Timber buildings range from semi-industrialized residential construction in North America, New Zealand and Japan to traditional structures built in many regions from bamboo, unprocessed logs and other woods. There is a wide variety of building types built from wood. In some areas, timber frames are infilled with masonry, to give the walls mass and thermal insulation. In others, walls are lightweight, or non-existant, to give good ventilation.

If worked in conjunction with planting regimes for the appropriate types of trees, timber and bamboo are sustainable building materials. Their inherent ductility, lightness and strength means that a well-built structure can withstand the lateral loads of earthquakes and high winds much better than other low-cost materials.

Timber and bamboo frames are capable of surviving considerable distortion before collapse, allowing quite severe foundation settlements and extreme loads to be tolerated before endangering the lives of the occupants.

However, timber and bamboo buildings depend critically for their structural performance on the strength of the connections: poorly built or decayed fixings can render

Traditional timber houses are resistant to earthquakes but are vulnerable to high winds. Philippines.

The resilience of timber makes it capable of undergoing considerable distortions before collapse. Timber house in an earthquake, Turkey.

buildings highly vulnerable to hazards. Being light, they are also more vulnerable to wind buffeting and being combustible, are always at risk from fire.

Timber and bamboo are organic materials and can rapidly decay if conditions allow it. If elements rot or are attacked by insects, the structure can become very weak. Metal fixings in wood lose their strength if the wood around them decays—a fact sometimes only discovered when the frame is subjected to unusual loads in natural hazards.

Timber and bamboo have great potential to provide very robust, resilient and hazard-resistant buildings, but, to achieve that, they must be carefully designed and built.

Why timber?

Before deciding to use timber, consider the advantages and disadvantages.

Advantages

The advantages of building in timber are:

- *High strength-to-weight ratio* — its strength-to-weight ratio makes it an excellent material for use in earthquake-resistant frames. Timber structures are much lighter than steel or concrete equivalents, which reduces the forces that have to be transmitted both vertically and horizontally
- *Flexibility* — the flexibility of timber structures enables them to withstand the ground shaking and wind buffeting with less damage. Joint flexibility makes the whole structure ductile
- *Easy to work with* — timber is light to transport, easy to work and assemble on-site
- *Energy saving* — much less fossil fuel is used in the production of timber-framed houses than if brick masonry, concrete or steel are used.
- *Local craft traditions* — well-established local craft traditions exist in all the major timber-growing parts of the world. Hazard-resistant construction can be developed by building on these traditions, and
- *Renewable resource* — all forms of timber can be grown in a sustainable fashion. However, a lot of timber used in construction of low-cost housing in the developing world is unfortunately not from sustainably growing sources.

Disadvantages

The disadvantages of building in timber are:

- *Increasing price and scarcity* — in many parts of the world the traditionally used types of timber are becoming scarce and expensive
- *Variable properties* — as a natural material, the properties of timber are variable, and defects occur. Where timber is stress-graded it can be used very efficiently, but elsewhere allowance has to be made for defects
- *Deterioration* — most timbers are subject to rot, fungal attack and insect attack in warm climatic conditions. Timber needs careful selection and design and many timbers also need preservative treatment to be durable
- *Combustibility* — the combustibility of timber is a potential problem: fires are an ever-present hazard in timber buildings, and
- *Limited range of sizes* — the limited range of sizes can be a problem for large span structures, but can be overcome by jointing and laminating.

For timber to be used successfully, six key principles should be applied:

1. Good foundations
2. Rigid vertical structure
3. Braced floors and roofs
4. Strong connections
5. Regular maintenance, and
6. Protect against fire.

These principles are elaborated in the following pages.

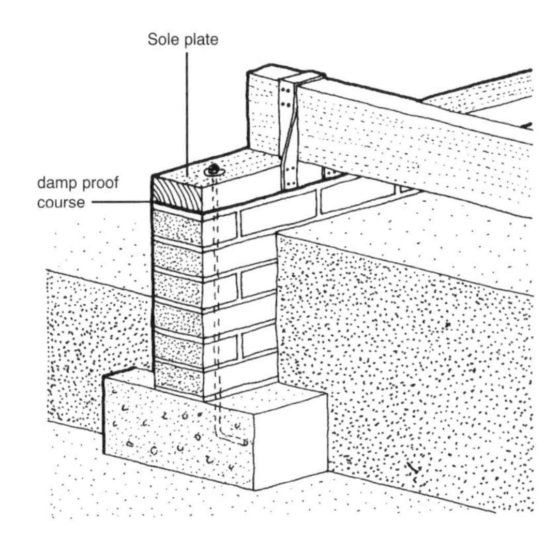

Figure 5.1 The sole plate should be firmly fixed to the foundations. Floor joists should be firmly fixed to the sole plate.

PRINCIPLE 1
Good foundations
Connect the building firmly to its foundation.

Foundations for timber structures have to carry much smaller forces than those for masonry and concrete buildings because the loads, both vertical and horizontal, are smaller. This means foundations can be simpler and cheaper.

However, the foundations must be designed to prevent major distortions of the building occurring. It is especially important to make positive connections between the building and its foundations.

Experience has shown that one of the most common causes of failure of timber buildings in earthquakes or winds is the separation of the building from its foundation by sliding, toppling or other failure of the connection.

Foundations may, according to loads and subsoil conditions, consist of:

- Separate concrete pad footings
- A strong concrete slab
- A reinforced or unreinforced concrete spread footing, and
- A stone or brick masonry spread footing.

For large buildings, in soft soil sites or in water, piled footings may be needed.

Whichever type of foundation is used, the connection of the superstructure to it is essential. The following principles should be followed:

- The timber superstructure should rest on a base member (sill plate or sole plate) which should be durable or treated wood and protected from damp
- Where the ground floor consists of a concrete slab, a damp proof course must be provided to protect the sole plate from moisture
- The strength and dimension of the sole plate should be sufficient to transmit to the foundations all the forces arising from gravity, wind and earthquake loads
- The sole plate should be firmly connected to the foundation using anchors, bolts or straps
- Where the ground floor consists of timber joists spanning between foundation walls, these should be strapped or anchored to the sole plate, and
- Where foundation walls (or piers) are raised to provide a ventilation space or crawl space below the floor, careful consideration must be given to the lateral forces they will have to resist. Cross-bracing will be needed. This can be done with diagonal boards or plywood sheeting (with some holes for ventilation).

Particular attention must be paid in cases of houses raised off the ground on posts in cyclone areas. Posts need to be properly anchored to the ground and braced between them, to prevent upward suction lifting an otherwise sound building off the ground.

PRINCIPLE 2
Rigid vertical structure
Provide well-braced walls in both directions.

The vertical structure transmits the gravity and lateral (earthquake or wind) forces acting

Well-braced walls provide resistance to earthquakes and high winds. Timber-framed house, Turkey.

on the roof and upper floors down to the level of the foundation. It has to be both strong and rigid and able to carry the forces acting in all directions. Thus the separate parts of the vertical structure — the main vertical and horizontal members — must be firmly tied together; and bracing must be provided to ensure lateral stability.

The essential principles of frame stability follow.

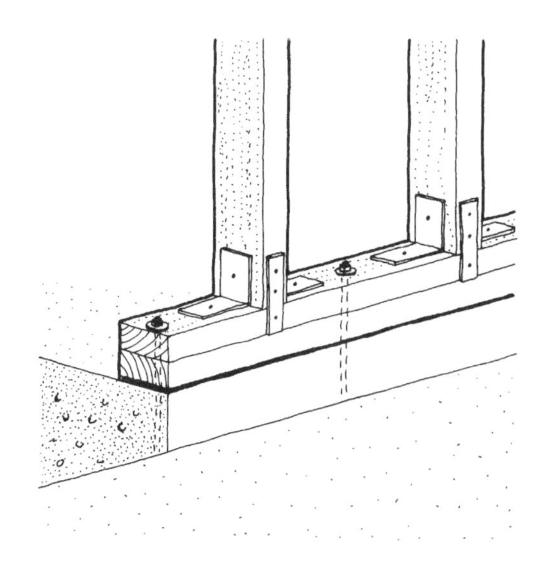

Figure 5.2 Vertical studs must be well-connected to top and bottom plates.

Wall construction:

- Provide regularly spaced walls in each direction
- Anchor the bottom of all wall panels firmly to the sole plate. This may involve bolting through the depth of the floor joists in the case of timber suspended floors
- Connect vertical studs to top and bottom plates by nailing, frame anchors or straps
- Provide horizontal blocking between vertical studs
- Tie the tops of wall panels together with a wall plate which should be bolted or strapped to the wall panel
- Wherever possible provide plywood sheathing to the outside face of the wall panel, nailed to each vertical and horizontal member at close centres. An alternative rigid sheathing is diagonal board sheathing — boards laid at 45° to the horizontal nailed to each vertical member
- If other methods of covering the frame are used (eg horizontal board, bamboo, plasterboard, chipboard) cross-bracing within the wall panels must be incorporated. This should consist of diagonal members set at 45-60° to the horizontal, notched into the frame by means of half joints, and
- Stiffen the corners of the frame at each storey by horizontal corner bracing, fixed in notches in the wall plate.

Openings:

- Openings should be limited to thirty per cent of the area of any wall
- Openings should be at least 500mm from corners, and spaced at 500mm from each other, and
- Panels between and above and below openings should be sheathed or cross-braced as described.

Wall cladding

In high winds and earthquake vibrations, timber frames flex and, if the wall materials are not firmly fixed to the frame, they will be shaken loose and could cause injury. Wall cladding should not be airtight in wind hazard areas. Moderate ventilation should be allowed to pre-

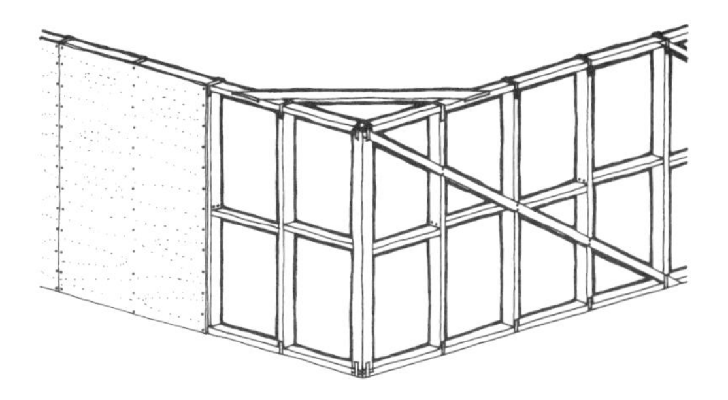

Figure 5.3 Brace the walls with horizontal blocking between studs and either cross bracing or plywood sheathing.

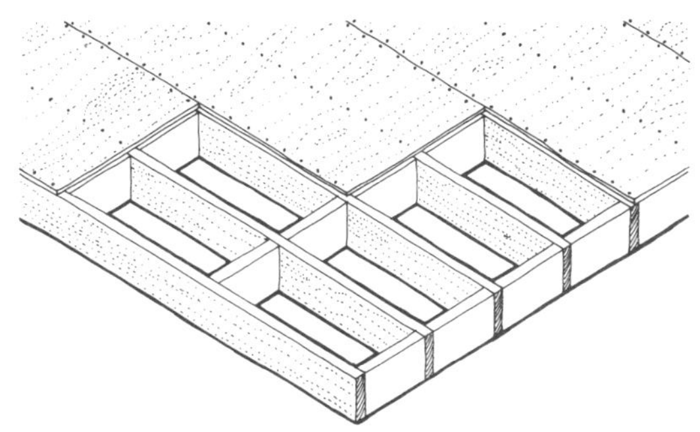

Figure 5.4 The floor should be well-braced with blocking pieces between the joists and a rigid floor covering fixed firmly to every joist.

vent large differences in air pressure building up between the inside and outside of the house.

- Insulation and wall cladding should be made from materials that are strong and should be securely fastened to the vertical structure, and
- Any metal fasteners fixing cladding to the frame should be galvanized, covered with protective paint or otherwise non-corrosive.

Masonry walls with timber frames

Timber structures with masonry walls should be designed to allow the timber frame to move separately from the masonry walls. Timber frames tend to flex under earthquake and wind loads, whereas masonry walls are stiff and stand firm. The net result is often damage being inflicted on the masonry by the timber frame. Masonry walls should stand independently of the frame and allow a movement space between frame and wall.

PRINCIPLE 3

Braced floors and roofs

Provide well-braced floor and roof construction.

The floors and roof are very important strengthening elements in frame construction. They should be designed to be as rigid as possible in their own plane to be most effective in transferring any lateral forces evenly into the vertical members of the frame.

The floor system of a typical timber building consists of:

- Floor joists, blockings and struts
- Metal straps and other connections, and
- Floor sheathing.

In earthquakes or high winds the horizontal structure of a timber building rarely fails, but failures are often caused by insufficient rigidity of the horizontal structure, or poor connections with the vertical frame.

Chapter 7 describes safe roof construction in some detail. The most important principles of floor construction follow.

Floor construction:

- Provide joist spacing and dimensions as required for normal vertical loading
- Ensure each joist is anchored to the wall plates on which it rests
- Provide timber blockings between joists over the supporting walls
- Provide cross-strutting between adjacent joists at regular centres (typically 2.5m) along the joist span
- Nail floor decking panels or boards to joists over the entire floor area at close centres, and to the edge blocking or blocking over internal walls
- The span to width ratio for a floor diaphragm should not exceed 3:1.
- Openings in the floor diaphragm (for staircases, chimneys etc) should be as few as possible. Larger openings should be positioned towards the centre rather than the

edge of the diaphragm, and
- All openings should be trimmed by joists or blockings, and the diaphragm nailed along all its edges.

PRINCIPLE 4
Strong connections

The foundations, frame, walls, floors and roof must be fixed firmly together to form a strong structural unit.

Many of the failures of timber buildings are caused by weak connections between the structural elements, rather than the disintegration of the elements themselves. The connections between the walls, floors and roof are areas subjected to substantial stresses when the frame flexes under wind, earthquake or other unusual load.

The integrity of the structural system relies on firm fixings, strong in tension and ductile: bolts, straps and nails can provide greatly increased connectivity and better structural safety at little extra cost.

Existing timber buildings can be strengthened much more easily than most other building types by fixing additional connectors to the structure without having to dismantle and rebuild. More straps, bolts and nails can be added to the structure while it is in position.

Strong connections are important:

- Between the foundations and the structural frame
- At corners, between the two stud walls or between wall and corner post
- Between walls and floors at each storey level, and
- Between the roof and the structural frame.

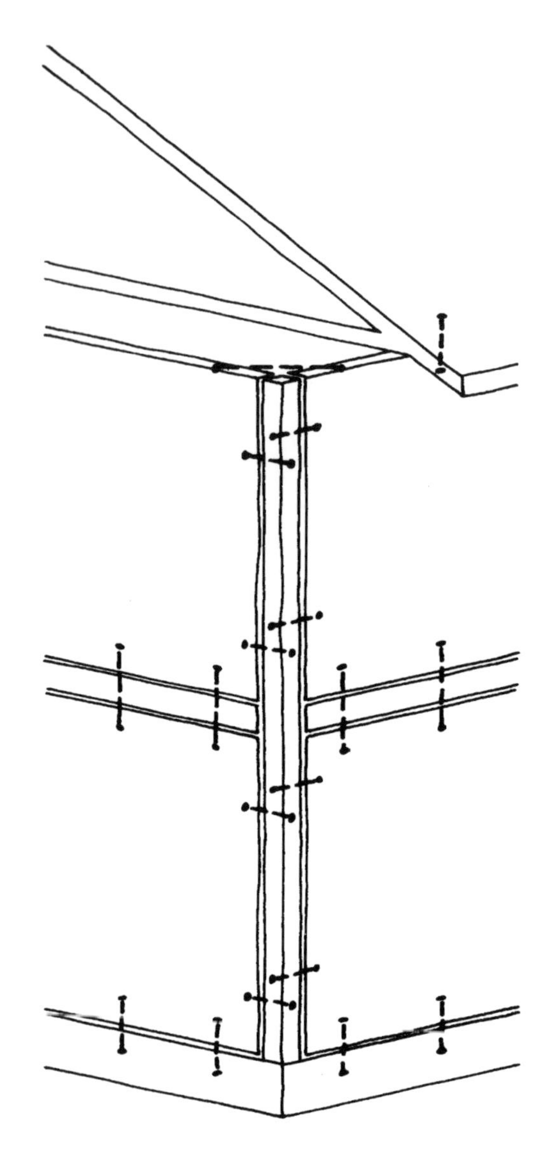

Figure 5.5 Strong connections between the elements of a timber frame are critical to its structural safety.

PRINCIPLE 5
Regular maintenance

Provide protection and maintenance throughout the life of the building.

A timber building needs protection and maintenance throughout its life if it is to survive the occasional stresses of earthquake or wind load.

One of the main problems of timber structures is their susceptibility to decay, insect attacks and their vulnerability to fire. Therefore, note that:

- It is of primary importance that the timber used is of good quality, well-seasoned and if possible treated with preservative. In some cases, preservative treatment may be needed to ensure durability, but pesticide treatments should be avoided where possible.

- Hardwoods are more durable than softwoods. Temporate hardwoods or sustainably managed tropical hardwoods can be used to build a longer-lasting structure
- The orientation and design should be such that the building will be protected from extreme exposure to prevailing winds, driving rain or direct sun that will hasten decay
- Care should be taken that water is not retained on surfaces nor enters the joints. Roof overhangs should provide protection for the timber structure from sun and rain
- A traditional measure to protect timber in contact with the ground is to use fire to char the ends (eg the bottom 30cm) of timber posts before burying them
- Protection from rot can be achieved by use of seasoned timber, but preservative treatment may still be needed
- Protect metal fixings from corrosion by using galvanized materials, anti-corrosive paint or other protective finish, and
- In termite areas, protect timbers from subterranean termites by using termite poison spread around the foundations. If available, seasoned and preservative-treated timbers will deter termite attack. Alternatively, buildings can be raised on posts which are provided with a metal termite shield. Regular inspection should be carried out to check for termites.

PRINCIPLE 6

Protect against fire

Reduce the risk of timber buildings burning.

One of the disadvantages of timber as a building material is its combustibility. In a major earthquake or other sudden hazard, fire sources are often created, for example the overturning of a stove or fractured fuel pipes. Building collapse greatly increases the probability of fire.

Some protective measures include:

- If available and affordable, timber should be treated with *fire retardant*. The fire retardant treatment is mostly effective on seasoned timbers. Some retardants are also preservatives
- Exposed timber surfaces (ceilings, walls, and floors) may be covered with plasterboard or other fire protective material
- Fire-resistant partitions, in masonry or other fire-resistant materials, should be introduced as fire-checks within roof or wall cavities to limit the spread of flame, particularly in large buildings or in terraces or adjacent buildings, and
- Areas where a fire is regularly used (kitchens, stoves, heaters) should be given special consideration. Hearths and fireproof surrounds can give a measure of protection. Fireplaces, lamps, stoves and other hazardous items should be secured so that they will not overturn or spill fuel if suddenly shaken.

In neighbourhoods of timber buildings, a major fire can rapidly spread from building to building. In a densely built area several ignitions may occur simultaneously and conflagration can become a major threat. The spread of fire in a neighbourhood is governed by building materials, proximity of buildings to each other, wind direction and velocity, availability of water supply and fire fighting tactics.

In building for safety, an important measure to reduce the risk of being caught in a neighbourhood fire is to maximize the distance between buildings. Timber buildings should be planned at low densities, with wide streets and space between structures. This may nevertheless be impossible in already built areas. In such cases *fire walls* between houses will help enormously. These can be simple cantilevered reinforced concrete walls of sufficient height and their location should take into consideration the direction of the prevailing winds in the area.

Summary

Make sure that:

- The timber frame is protected from ground moisture
- The frame is firmly connected to the foundations
- The building has equal amount of walls running in perpendicular directions
- The frame is sheathed in plywood or diagonal boarding or cross-braced
- The floor is covered with a sheeting material, firmly nailed to all floor joists
- Rafters are tied to the walls
- Roof coverings are held in place, and
- All timber is well-maintained.

Avoid:

- Unbraced cripple wall construction
- Unbraced raised floors
- Unventilated timber floors
- Sole plates which are not protected from moisture
- Unbraced foundations posts or walls
- Large openings on external walls
- Large openings in the corner of the floor, and
- Unsecured stoves and fireplaces.

Further reading on building safely in timber

Australian Overseas Disaster Response Organization (1988). *Disaster-resistant Construction for Traditional Bush Houses: a handbook of guidelines*. Available from: AODRO, Suite 201, 2nd Floor, 381-383 Pitt Street, Sydney, New South Wales 2000, Australia.

Applied Technology Council (1980). *The Home Builder's Guide for Earthquake Design* (condensed version of ATC-4). Available from: ATC, 3 Twin Dolphin Drive, Suite 275, Redwood City, Calif. 94065, USA.

Eaton, K.J. (1980). *How to Make your Building Withstand Strong Earthquakes*. Available from: Building Research Establishment, Garston, Watford WD2 7JR, UK.

Mayo, A. (1988). *Cyclone-resistant Houses for Developing Countries*. Available from: Intermediate Technology Publications, Myson House, Railway Terrace, Rugby, CV21 3HT, UK.

Schilderman, T. (1990). 'Earthquake protection for poor people's housing'. *Appropriate Technology*, Vol. 17, No.1, June 1990. Available from: Intermediate Technology Publications.

United Nations Industrial Development Organisation (1985). *Timber Engineering for Developing Countries* (Part 1 to 5). Available from: UNIDO, Vienna, Austria (Public. No. 10.606 to 10.610).

Chapter 6 Building Safely in Reinforced Concrete

Reinforced concrete and hazards

Reinforced concrete (RC) is today the most popular building material worldwide for buildings of more than three storeys. It is made from materials which are available almost everywhere, and its basic properties are reasonably well-understood by both engineers and builders.

However, in recent years reinforced concrete buildings have made headlines in news from all over the world when spectacular collapses have occurred, either due to surcharge loads or inadequate design and other causes apart from natural hazards.

The experience of several recent earthquakes has also given cause for concern about the safety of reinforced concrete buildings built in recent years. In the earthquakes in Bucharest (1977), Southern Italy (1980), Greece (1981, 1986), Mexico City (1985), El Salvador (1986), Armenia (1988), Philippines (1990), eastern Turkey and Egypt (1992), many multi-storey buildings collapsed because they were not designed and constructed to resist the shaking which they experienced. Flash floods are also destructive to RC buildings mainly by washing away buildings with shallow foundations. In case of high winds damage tends to affect non-structural elements, such as parapets, window opening, external additions especially in roofs.

In spite of its universality and great popularity, it is important to understand that designing

Reinforced concrete structures can be highly vulnerable to earthquakes and land instabilities if not designed well. Irregular building with weak ground floor, collapsed in an earthquake, Turkey.

and building a safe reinforced concrete structure in an earthquake area is not an easy matter. A code of practice appropriate to the place where the building is to be built should be followed. Codes of practice are complex documents containing many detailed requirements and recommendations. They need interpretation by an experienced engineer.

The most common type is the *reinforced concrete frame with masonry infill walls.* The frame consists of reinforced concrete columns and beams and the most usual form of flooring is a cast *in-situ* reinforced concrete slab, with continuous reinforced concrete footings. The infill masonry is unreinforced or reinforced with vertical and (or) horizontal reinforcement, but usually not designed to carry any load and mostly made with concrete blocks or clay bricks.

The following principles apply to buildings in earthquake areas. But buildings constructed according to them will also be resistant to other hazards.

The first rule of building safely in reinforced concrete is:

In earthquake areas do not build in reinforced concrete except under the supervision of a structural engineer.

The recommendations and principles that follow are intended to offer guidance on the planning and form of buildings to help avoid some of the most common mistakes which can lead to unsafe buildings. They do not replace the detailed requirements of codes of practice and the knowledge and experience of an engineer.

Why concrete?

Before deciding to build in reinforced concrete, carefully consider both the advantages and disadvantages.

Advantages:

- Materials for reinforced concrete construction like aggregates, cement and steel reinforcement are now widely available in most parts of the world
- Building in reinforced concrete need not be capital-intensive. Relatively large buildings can be built with a small investment in plant and equipment
- Well-designed and well-built concrete can have excellent earthquake resistance even when used for multi-storey buildings
- Reinforced concrete offers (even within the limitations imposed by design for

earthquake resistance) the potential for great flexibility in planning
- Reinforced concrete is a highly durable material — if well-made it will retain its initial strength without requiring maintenance for decades, and
- Reinforced concrete is a fire-resistant material.

Disadvantages:

- Good-quality construction depends on good workmanship and supervision on site even more than when using the other common materials
- Defects in concrete and reinforcement are difficult to see after casting, so subsequent inspection cannot easily reveal serious weaknesses due to poor design or construction
- Good earthquake performance depends on the existence and application of good building design codes for earthquake loads — in many countries code development and enforcement is very weak
- Concrete structures are heavy and often large; if they do collapse, a great loss of life is likely to occur
- The strength of the whole structure may depend on the performance of a few critical parts
- In some areas, the materials needed for good-quality construction, such as sand for fine aggregate, timber for formwork, are becoming scarce, and
- Reinforced concrete buildings cannot easily be altered without damaging their structural integrity.

The key principles of building safely in reinforced concrete are:

1. Robust building form
2. Regular frame
3. Attention to element design, and
4. High-quality construction.

PRINCIPLE 1

Robust building form

Make the building symmetrical and simple both in plan and elevation.

The studies of damage from past earthquakes show that the types of reinforced concrete buildings which perform best in earthquakes are the simple, regular and symmetrical ones.

Buildings which perform worst are those which have:

- Complicated plan shapes
- Long and thin plan shapes
- Overhangs and setbacks between one storey and the next
- Open ground floors used for shops, storage or car park
- Asymmetrical arrangement of walls, staircases or columns at each floor
- Buildings adjacent to each other without separation to avoid pounding during the earthquake shaking, especially when they are of different height and size or when floors are not at the same level, and
- Corner buildings with excessive openings in front and no openings in sides, especially when built in contact to adjacent structures without separation.
- Multi-storey buildings on very soft ground.

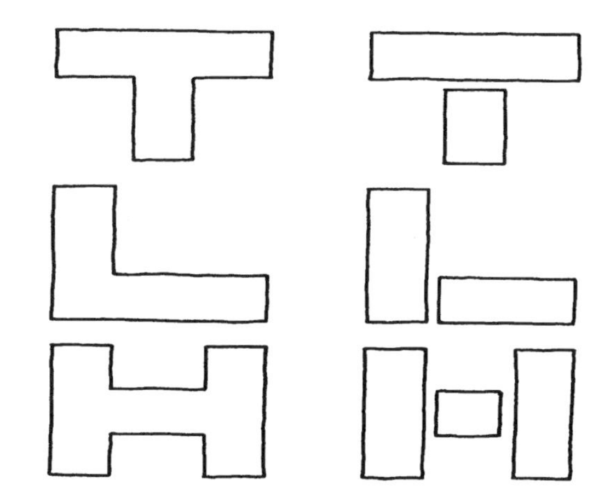

Figure 6.1 Separate complex plan shapes into structurally independent units.

Figure 6.2a
Bad construction practice.
On a sloping site, stepped foundations like this are unstable. Avoid this type of foundation..

Figure 6.2b
Good construction practice.
On a sloping site, it is better to terrace and have flat foundations.

Reinforced concrete buildings more than five storeys in height have often performed worse than low-rise buildings.

Reinforced concrete buildings can also be badly hit by earthquakes if the soil conditions under the foundations vary from one end of the site to the other, or if built into steeply sloping ground.

General principles:

- The overall building plan should be of simple and symmetrical shape
- T-, L- and H-shaped plans, and complex plans with re-entrant corners should be avoided, or separated into structurally independent rectangular units
- The building should be not more than 40m long, unless separated into different independent structures
- Structurally independent parts of the building should be separated by vertical gaps not less than 100mm (150mm for three storeys)
- The foundations and subsoil conditions should not vary significantly across the site
- Building on steeply sloping sites should be avoided, and
- For buildings higher than five storeys, particular attention should be given to the possible effect of the subsoil on ground shaking.

PRINCIPLE 2

Regular frame

Design a continuous, regular reinforced concrete frame.

The simplicity and regularity needed in the overall building form has to be carried through into the detailed design of the building frame and the way it is reinforced if it is to survive the next earthquake. Experience has shown that a structure has the best chance of surviving an earthquake if:

- The load-carrying members are uniformly distributed
- The columns and walls are continuous from floor to roof
- All the centre lines of columns and beams meet each other
- Columns and beams are nearly the same

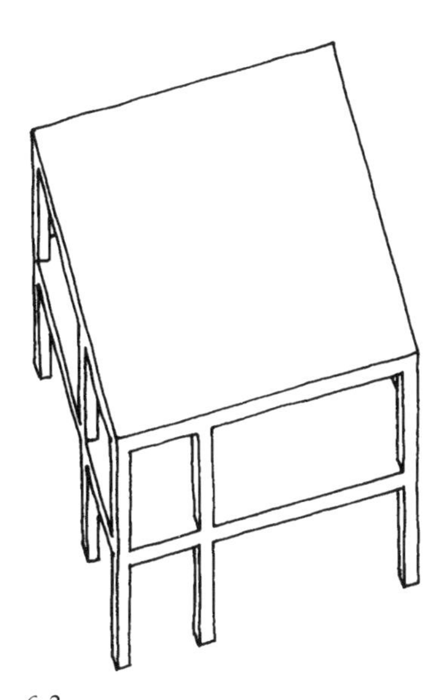

Figure 6.3a
Bad design practice.
Irregular frame spacing creates uneven loads and results in a vulnerable building.

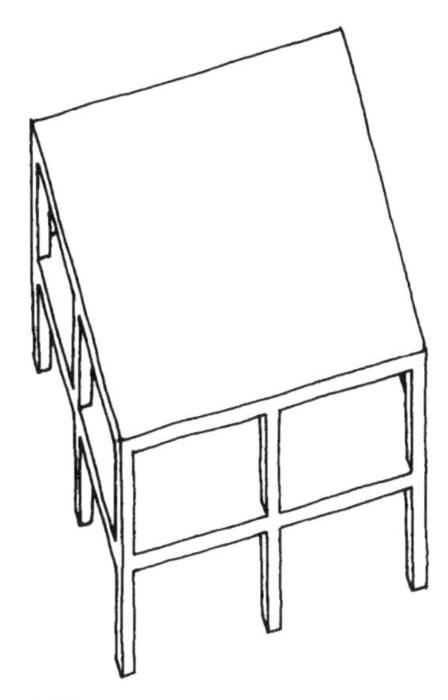

Figure 6.3b
Good design practice.
A regular and symmetrical spacing of the frame allows equal distribution of loads and results in a safer structure.

width
- No principal members have sudden changes in cross-section, and
- The structure is as continuous and monolithic as possible.

The structure should be designed with many alternative ways of transmitting the horizontal loads to the ground. A structure designed so that if one element fails others can take its load will have a better chance of surviving an earthquake than one where failure of the whole structure would follow from failure of one member. Special attention to reinforcement details is essential if rapid disintegration of joints in earthquakes is to be avoided.

Structural points to remember are:

- Within the overall plan, the arrangement of vertical supporting members should be regular and symmetrical
- The vertical structure should be continuous from top to bottom of the building, without changes from one floor to the next
- Floors should be at the same level across the whole building, not stepped
- Masonry or other structural infill within the structural frame should be continuous from top to bottom without major differences from floor to floor, and
- The position and area of openings should not vary significantly from floor to floor.

Columns

- Columns should be uniformly distributed in plan, each carrying approximately the same amount of load. A good rule is that the area of floor carried by one column should not be more than $40m^2$
- Columns and walls should be in straight lines in both directions
- The frame should be strong in both directions and the spans in one direction should not be more than twice those in the other direction, and
- Columns should be stiffer than the beams that frame into them.

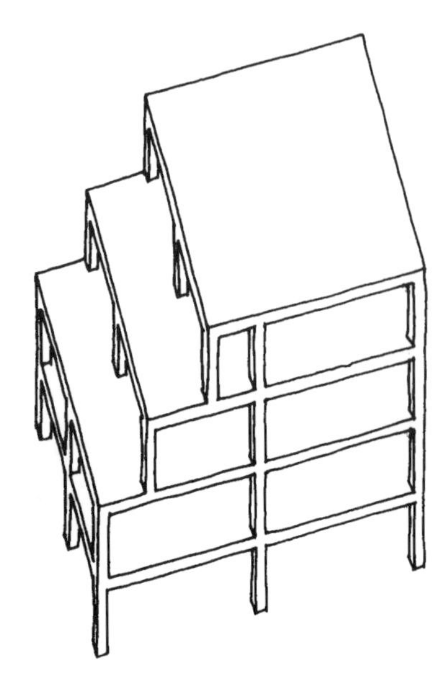

Figure 6.4a
Bad design practice.
An irregular frame is unbalanced and weaker against unexpected or hazardous loads.

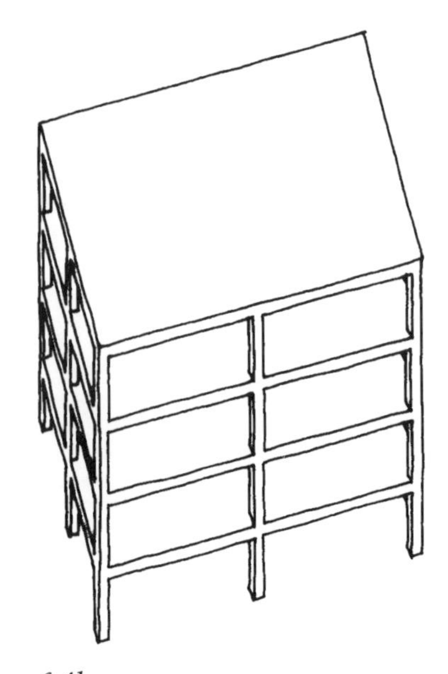

Figure 6.4b
Good design practice.
The safest structure is one in which the frame is regular vertically as well as horizontally.

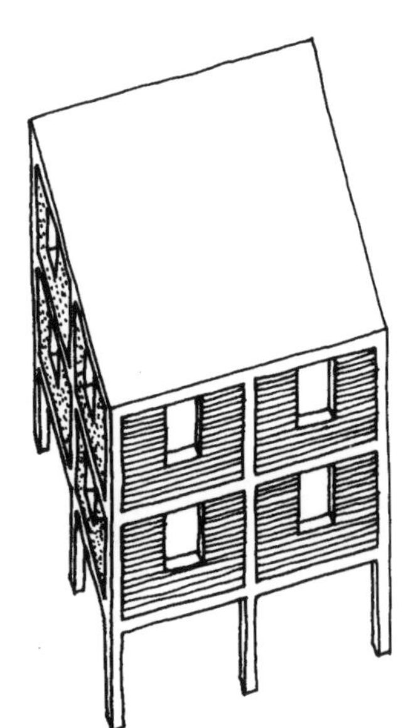

Figure 6.5a
Bad design practice.
Weak, open ground floor.

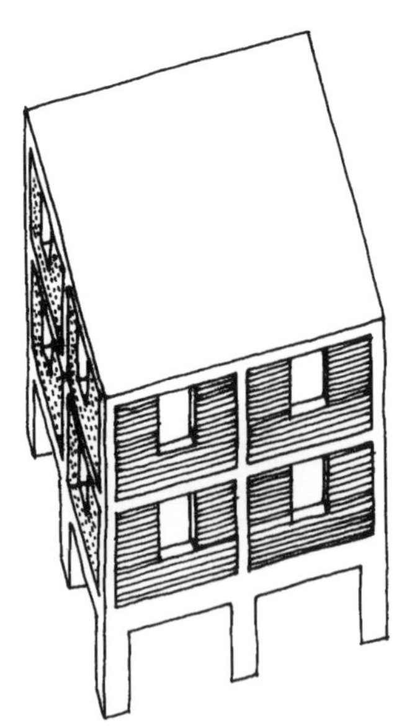

Figure 6.5b
Better design practice.
Stiff ground floor columns.

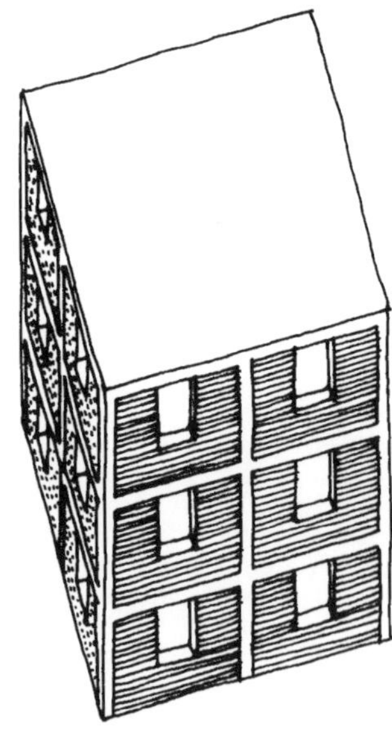

Figure 6.5c
Good design practice.
Solid structure on all floors.

Continuity of infill walls from one storey to another avoids differences in stiffness and the creation of potential weaknesses.

Figure 6.6 Columns should be stiffer than the beams that frame into them.

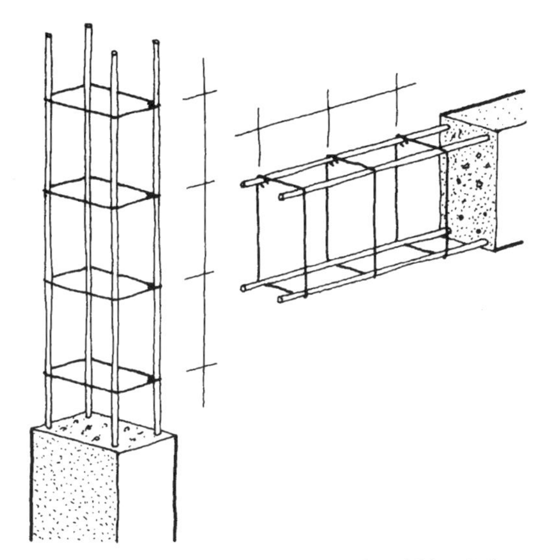

Figure 6.8 Reinforcement cages should be tied with hoops at regular intervals.

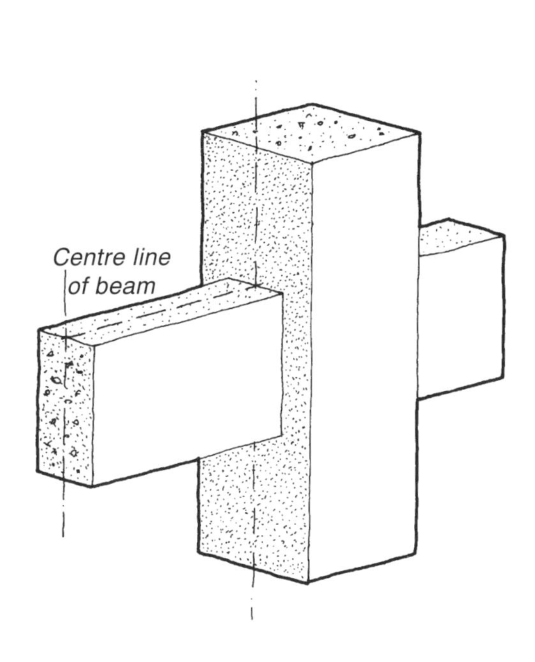

Figure 6.7 Centre lines of beams should intersect with centre lines of columns.

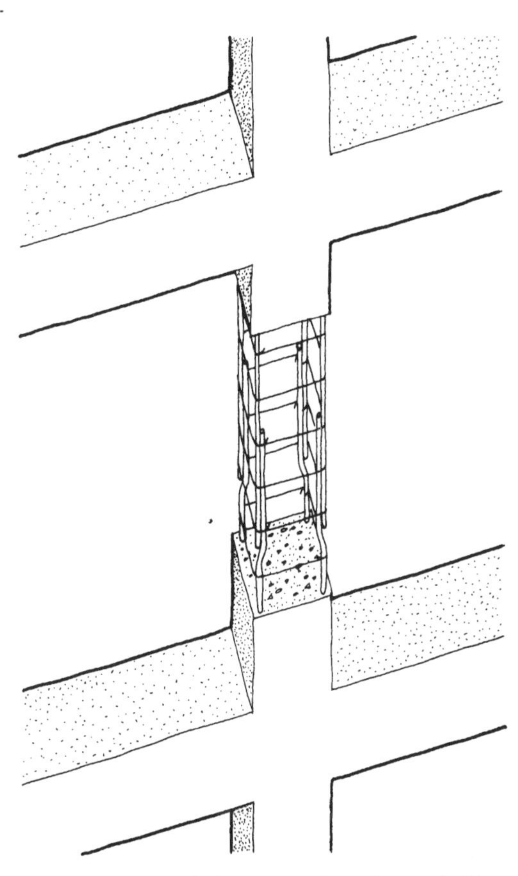

Figure 6.9 Lap reinforcement in columns half-way up rather than near junctions with beams.

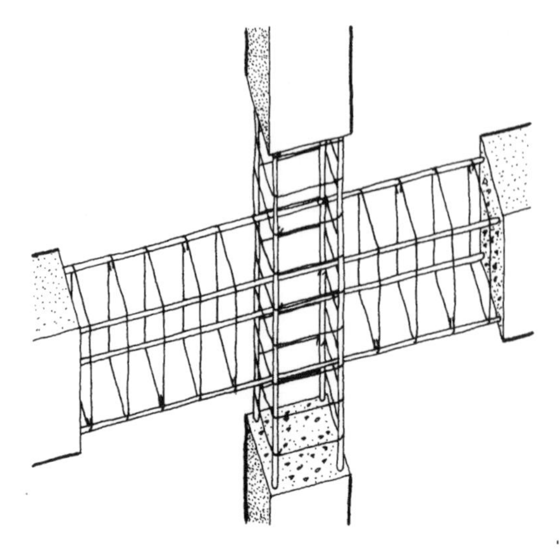

Figure 6.10 Carry the reinforcement straight through junctions.

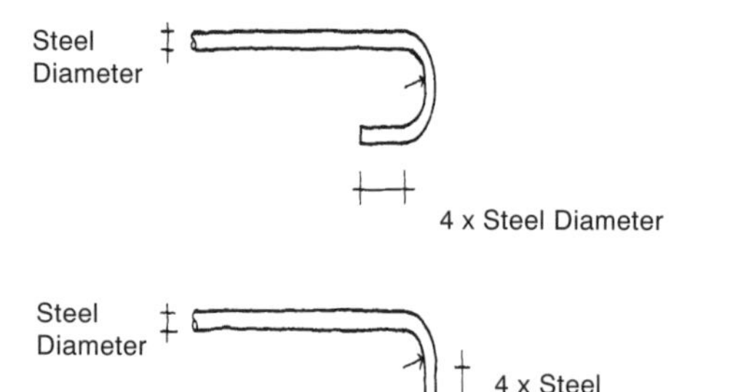

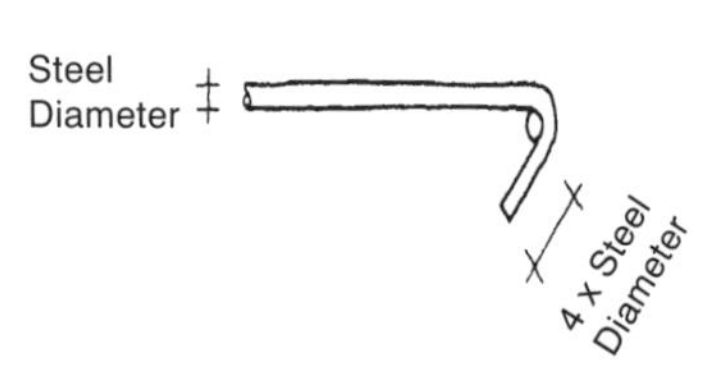

Figure 6.11 Reinforcement hooks allow the steel to be tied and firmly anchored at its end.

Beams

- Centre lines of beams should intersect with centre lines of columns
- The widths of beams and the columns which support them should be about the same
- Grids of columns and beams should, wherever possible, be laid out at right angles to each other
- A good rule is that beam spans should be not less than 3m or more than 7m
- Beams and columns should, wherever possible, be cast simultaneously in one piece, to form a single continuous element, and
- Flat slab floors (without beams) should be avoided unless column dimensions are large.

Reinforcement:

- Reinforcement should be of deformed steel wherever possible
- Minimum cover to reinforcement should be maintained everywhere (typically 50mm in foundations, 35mm in internal columns and 25mm in slabs)
- Reinforcement cages in columns and beams should be tied with hoops at regular intervals. These transversal hoops should be no more than 15cm from each other along the height of the column and the length of the beam
- Vertical reinforcement of columns should run continuously through connections and lap at mid-height
- All beams should have longitudinal reinforcement at both top and bottom; all longitudinal reinforcement should be continuous through connections
- Columns should have transverse stirrups (hoops) or spiral reinforcement over their full height, not less than 10mm in diameter, with closer spacing at top and bottom
- Particular attention should be paid to code requirements for minimum lap splices in both columns and beams
- The ends of all reinforcing bars including stirrups should be anchored properly
- The steel bars must be shaped to a hook form at the ends so that they can be well-anchored. Plain bars must have U-shaped hooks, for deformed bars L-shaped hooks are sufficient. For bars thinner than 12mm, 135° hooks are recommended. Transversal reinforcement should also terminate with hook
- Reinforcement congestion at connections should be avoided (minimum spacing between bars should be greater than max-

imum aggregate size). If the reinforcement seems congested, larger members should be used, and

- Reinforcement must be continuous from foundations into the column bases.

PRINCIPLE 3

Attention to element design

Slabs, foundations and infill walls need special consideration.

Because everything is connected the performance of a reinforced concrete building as a whole depends on the good performance of all parts of the structure, not just the main reinforced concrete frame. Failure of elements can be local failures, which endanger life, or they can trigger the complete failure of the building. These elements need particular attention. Some guidelines follow.

Slabs and cantilevers

- Slabs should have a minimum thickness to control deflection
- Slabs should have extra reinforcement in areas of complex shape (near openings, corners) or near vertical supports
- Reinforcement should be regularly spaced in each direction and in both top and bottom of the slab (minimum spacing 200mm)
- Even if not required for strength, a minimum amount of reinforcement should be provided to control shrinkage (0.25 per cent of the cross-sectional area is typical)
- In slabs and cantilevers it is very important that the top reinforcing bars are maintained in their correct position, throughout the construction and until concrete is being cast. Builders must take care not to step on them
- In cantilevers the reinforcement must not be allowed to bend before and during casting
- If precast slab units are used, special precautions (outside the scope of this book) are needed for creating continuity, and
- Slabs should be cast along with columns and beams as one continuous operation.

Foundations:

- Foundations may be individual footings, piers or piled foundations or other types according to loads and soil conditions
- The same type of foundation should be used under the whole building
- Individual foundations should be connected with tie beams
- Foundations of buildings with unbraced frames should be connected with stiff tie-beams, and
- In areas where liquefaction is a potential problem, foundations have to be specially constructed, for example by piling to below the layer of weak soil that could liquefy. Alternatively various methods of soil compaction can be applied, especially in case of an area where many buildings are to be built. However, both these methods involve significant extra cost and it may be more prudent to search for an alternative site.

Infill walls

- Infill walls should be arranged as uniformly as possible on every floor
- The use of different masonry materials or mortar mixes in different parts of the building should be avoided
- If the infill panel is structurally separated from the frame to allow relative movement during earthquakes it should be tied to the frame with flexible ties to prevent it from falling out, and
- The infill panels may be treated as part of the structure; their strength and the strength of the surrounding structure must then be adequate to resist the local forces. Horizontal reinforcement in the courses or half-height reinforced concrete bands should be used.

PRINCIPLE 4

High-quality Construction

Follow established good construction practice.

Reinforced concrete is a safe material for building in earthquake areas only if it is constructed

Figure 6.12a
Bad construction practice.
Infill panels that are not reinforced or tied into the structure will be badly damaged when the frame moves under strong loads.

Figure 6.12b
Good construction practice.
Infill panels are part of the structure: either reinforcement or concrete bands will provide horizontal ties into the structure and reduce damage from earthquakes and high winds.

to a high standard. Experience shows that many of the failures of individual reinforced concrete buildings in past earthquakes have been the result not of design failings, but of failure to implement the design.

A virtue of concrete is that, because it is cast on-site (*in situ*), a concrete frame can be made as one single piece, with continuity of reinforcement through every connection. But, as it is made on-site, the quality of reinforced concrete depends heavily on the experience of the craftsperson and on the quality of the supervision of the work.

Reinforced concrete is a composite material in which both concrete and steel play important roles. Concrete is a brittle material, cracking easily in tension, and useless for building unless reinforced. But the reinforcement needs the strength of the concrete in compression, and it needs protection from fire and corrosion. This intimate interaction between concrete and steel can easily break down unless the process of making reinforced concrete is carried out to a high standard.

The following aspects of the work need to be planned and carried out carefully: selecting and storing materials; building formwork; fixing reinforcement; mixing and placing concrete, curing concrete. Some of the most important points to remember about each follow.

Selecting and storing materials

- Aggregates can come from many different sources. For fine aggregates (less than about 5mm diameter) crushed rock or river sand sources are preferable; for coarse aggregates (greater than 1mm) crushed rock or gravels are most suitable. But many waste and man-made minerals are also suitable
- Aggregates should always be clean (free from mud or fine dust) and have a well-controlled distribution of particle sizes
- Reinforcement should be supplied according to recognized standards; it should be stored on-site to protect it from contamination. All loose rust should be removed
- Cement should satisfy national standards. It should be protected from damp until used, and all affected material discarded

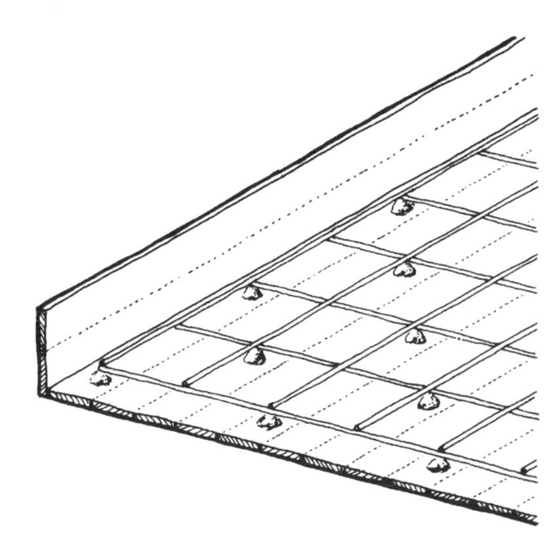

Figure 6.13 Spacers should be used to keep reinforcement in position while concrete is being poured and to maintain a minimum cover of concrete.

- Aggregates should be stored on-site to protect them from contamination by soil and from mixing with each other
- Water should be from an uncontaminated source, preferably drinkable, and
- Ready-mix concrete from a reputable source of supply should be used where possible.

Building formwork

- Formwork and its supporting framework should be strong enough to resist sagging or deformation during casting
- Formwork should be watertight, constructed to prevent leaking of water or cement paste during casting either between formwork joints or at junctions with existing work, and
- Temporary vertical forms should be provided at all daywork joints.

Fixing reinforcement:

- Reinforcement should be preformed in rigid cages or mats
- Bending of bars on site should be done using a former
- Reinforcement should be held in position in relationship to the formwork (by ties, spacers, etc), and

Steel reinforcement can be bent using a former on site. Construction project, Turkey.

- Minimum specified cover to reinforcement should be maintained everywhere by non-corroding spacers.

Mixing and placing concrete:

- Concrete should be mixed by weight according to mix ratios which are appropriate for the aggregates used, and the strength and workability needed
- Mixing (whether by hand or using a concrete mixer) must be thorough
- The concrete should be placed immediately after mixing
- The mix should be designed so that segregation of fine and coarse material does not occur during transportation, and
- After placing, the concrete must be compacted using a mechanical vibrator, so that all air is removed.

Curing:

- In cold climates, the concrete must be protected from frost after casting
- In hot climates, the surface must be covered after casting to prevent loss of liquid
- After removal of formwork, the concrete should be protected from loss of moisture for a minimum of two weeks, and
- Any areas of badly compacted or honeycombed concrete should be immediately cut out and recast in concrete.

Summary

Make sure that:

- In earthquake areas the design and construction supervision are done by an experienced structural engineer
- The building layout is as symmetrical as possible
- The soil conditions under the site are considered when deciding the height of the building
- Columns are continuous and uniformly distributed
- Beams and column lines intersect
- Good concrete construction practice is followed
- Reinforcement has proper size, spacing, cover, splicing and bending. Deformed steel bars are preferred to round bars
- Extra reinforcement is allowed around openings and vertical supports
- Infill walls are not very heavy and are isolated from the frame or reinforced with vertical and horizontal steel bars, and
- Adjacent buildings are spaced to avoid pounding.

Avoid:

- Complex plan shapes
- Recessed elevation
- Large openings in corner buildings
- Soft strong buildings
- Large masses located on roofs
- Using aymmetrical infill wall layout
- Large cantilevered overhangs or recesses
- Eccentrically placed reinforced concrete cores for lift shafts and stair wells
- Split-level floors
- Eccentrically placed large roof appendages like water tanks
- Alterations to the structure like knocking off infill walls or columns, and
- Building on liquefiable soils.

Further reading on building safely in reinforced concrete

Dowrick, D.J. (1987). *Earthquake-resistant Design for Engineers and Architects,* John Wiley and Sons, Chichester.

IAEE (1988). *Earthquake-resistant Regulations — a World List,* International Association for Earthquake Engineering, Tokyo.

Lagorio, H.J (1990). *An Architect's Guide to Nonstructural Seismic Hazards,* John Wiley and Sons, Chichester.

Spence, R.J.S., and Cook, D.J. (1983). *Building Materials in Developing Countries,* John Wiley.

UNDP / UNIDO (1984). 'Building Construction under Seismic Conditions in the Balkan Region', Vol. 2 *Design and Construction of Reinforced Concrete Buildings,* UNIDO, Vienna.

Chapter 7 Building a Safe Roof

Roofs and hazards

Roofs and earthquakes

Roofs are an essential part of the building structure. Although not the primary issue of concern in earthquake-resistant design, roofs must be designed to resist the lateral forces and add to the structural integrity of a building. In past earthquakes buildings were damaged solely because of a mistake in the design of the roof. It is essential that the roof is securely tied to the vertical structure of a building. The roof should have limited openings and discontinuities in order to act as a diaphragm. Light roofs are preferable. Make sure the roof is not overloaded. Eccentrically placed roof appendages should be avoided.

Roofs and volcanoes

For buildings situated in areas near active volcanoes consideration must be given to the possibility of having impact with volcanic eruption material or being overloaded with ashfall deposits. In major eruptions even settlements more than 50km away might experience ashfall of 5cm or more. To resist ashfall loads it is best if the roof spans are narrower and rafters are provided for longitudinal support. Roof overhangs and weak vertical structure are other possible vulnerable parts in case of excessive loading.

Roofs and strong winds

The most important safety issue for buildings in cyclone areas is that related to the roof struc-

Ashfall from volcanos or heavy snow loads can add unexpected weight to the roof structure and cause considerable damage. Damage from heavy volcanic ashfall, Philippines.

ture. The cost of making a house more resistant to cyclones will not add more than fifteen per cent to its final cost. Therefore, it makes very good economic sense in countries with frequent windstorms. The principles to be addressed are:

1. Aerodynamic roof form
2. Roof connected to structure
3. Well-fixed roof cover, and
4. Regular maintenance.

The following sections consist of brief discussion about each of these principles. These apply mostly to buildings in windstorm areas but will contribute to the earthquake resistance as well.

PRINCIPLE 1

Aerodynamic roof form

Design the form of the roof to reduce the wind loads on it.

The most important guidelines that you must follow are:

- Roof pitches should be around 30° to 40° to reduce the effect of suction and uplift forces. Avoid pitches of around 5° to 10°

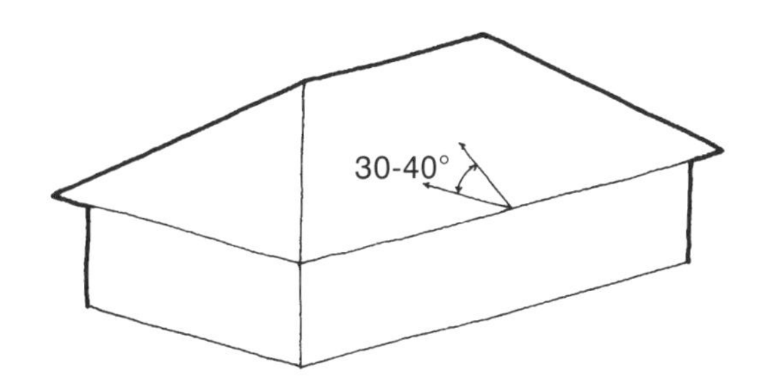

Figure 7.1 A hipped roof is less vulnerable to high winds than a gable roof. The best pitch to reduce vulnerability to wind is 30-40°.

- Avoid sharp edges in the shape of your roof. Prefer a *hipped roof* instead of a roof with gable wall, because the gable end walls are vulnerable to suction or wind pressure. Otherwise the gable ends should be firmly tied to the rest of the structure
- Avoid overhangs that are more than 0.8m from the external wall. Keep rafter notching to no more than twenty-five per cent of its cross-section
- Provide vents in the roof to relieve the upward pressures during the storm. This is of particular use to existing buildings that may have any of the previous weaknesses
- A masonry or concrete parapet around the roof edges can reduce the suction forces, but make sure that is firmly tied with the rest of the structure, and
- If affordable a *flat RC slab roof* provides superior protection as long as the rainwater is properly drained and thermal insulation provided. However, because such a slab is heavy in earthquake areas it needs adequately braced vertical structure.

PRINCIPLE 2

Roof connected to structure

Tie the roof firmly to the supporting structure.

- Provide rafters at spacing and dimensions as required for normal vertical loading

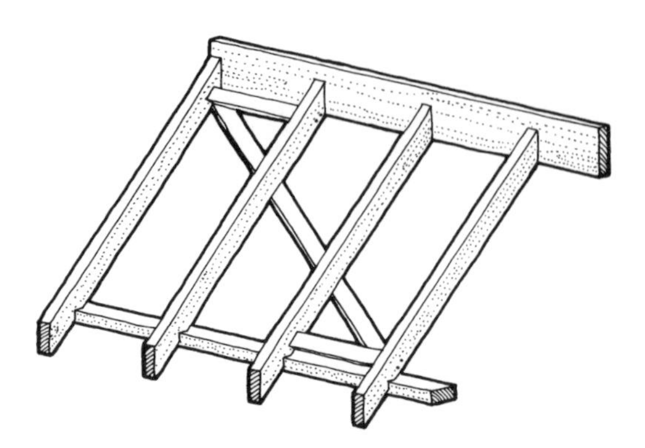

Figure 7.2 Cross-brace the roof to tie it into an integral structural unit.

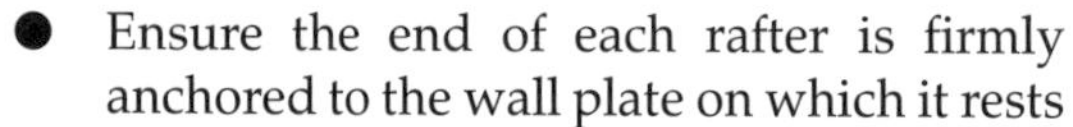

- Ensure the end of each rafter is firmly anchored to the wall plate on which it rests
- Provide cross-bracing or a rigid diaphragm in both the plane of the roof and the ceiling
- The roof covering material should be as light as possible (fibre cement and sheet roofs rather than clay tiles)
- Tie all roof covering materials firmly to their supports to prevent dislodgement in earthquakes
- Openings in the roof structure should be as few as possible, and cross-braced
- Tie the roof firmly to the walls, to prevent it being torn apart or blown off
- The connection between roof and vertical structure must be firm. Prefer if possible the use of metal straps or through bolts with washers at both ends, instead of simple nails, and
- The roof structure should be strengthened with trusses, collar-ties and corner bracing.

PRINCIPLE 3

Well-fixed roof covering

The roofing materials must be firmly fastened to the roof structure.

Loose roofing materials are a very common cause of damage during storms. Apart from leaving the roof exposed they can be blown long distances and cause injury and damage to

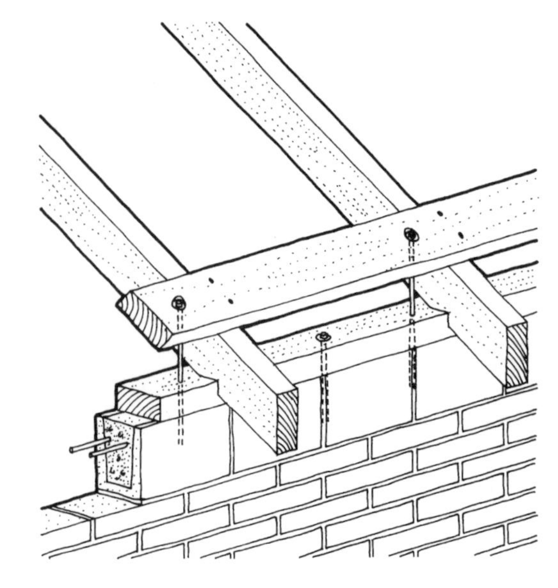

Figure 7.3 Anchor the roof very firmly to the structure supporting it.

other buildings. The most important guidelines are:

- Corrugated iron sheets must be frequently nailed or bolted to the truss
- Tiles, slates etc, must be fastened individually. Transversal masonry or cement mortar ribs can be built along the roof every 1.5 to 2m to keep the tiles together, and
- Use of loose stones to weigh down tiles should be avoided.

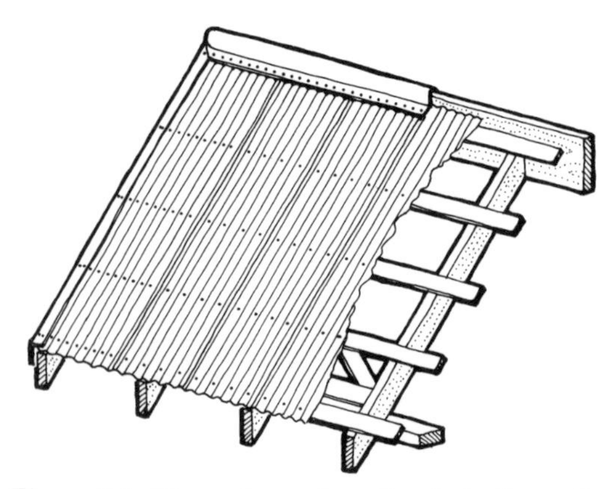

Figure 7.4 Fix roof coverings firmly to the roof structure.

Flat roof coverings can add considerable horizontal stiffness to a structure if cross-braced or covered with plywood sheeting. Roof construction in earthquake reconstruction project, Yemen.

PRINCIPLE 4

Regular maintenance

Roof maintenance may save your building in a storm.

- Before a cyclone strikes, make regular checks on the structural health of the roof. Check especially around the corners and along the ridge
- Replace any damaged or fallen tiles
- Fix down any roofing sheets that have begun to come loose
- Replace any corroded metal tile clips, nails or bolts used in the roof or to fix coverings
- Make sure that your roof has not been weakened by any other effects like rotting or insect attack, and
- Secure aerials or chimneys with wires, straps or bolts. Make sure the fastenings are corrosion-resistant.

Summary

Make sure that:

- The roof is firmly tied to the walls or other vertical structure
- The roof cover is firmly fastened to the roof structure
- The roof is ventilated to relieve upward pressure during the storm
- The roof is regularly maintained, and
- The roof acts as a diaphragm, having no irregularities in its shape and limited openings.

Avoid:

- Low pitch roofs or sharp edges
- Gable walls
- Excessive roof openings or roof discontinuities
- Excessive overhangs, and
- Large masses located on roofs.

Further reading on building a safe roof

Diacon, D. (1992). *Typhoon-resistant Housing in the Philippines — A Success Story.* Information from: Building and Social Housing Foundation, Memorial Square, Coalville, Leicestershire LE6 4EU, UK.

Eaton, K.E. (1980). *How to Make Your Building Withstand Strong Winds*, BRE Public.

Eaton, K.E. (1981). *Buildings and Tropical Windstorms*, BRE Overseas Building Notes, 188.

Eaton, K.E. and Reardon, G. (1985). *Cyclone Housing in Tonga*, BRE Public.

Greenwood, R.F. (1992). *Hurricane-resistant Construction.* Available from: The Reporter Press, 147 West Street, Belize City, Belize, Central America.

Mayo, A. (1988). *Cyclone-resistant Housing for Developing Countries*, BRE publication distributed by Intermediate Technology Publications.

Norton, J. G. Chantry, and Ngyen Si Vien (1990). *Typhoon-resistant Building in Vietnam*, in *Mimar* No. 37.

Shri Reddy, I.A.S. (1992). *Building for Safety: a case study of cyclone-prone coastal region of the eastern state of Andhra Pradesh, India.* Information from: National Institute of Rural Development, Rajendranagar, Hyderabad-500 030, India.